Biology of Nematodes

TERTIARY LEVEL BIOLOGY

A series covering selected areas of biology at advanced
undergraduate level. While designed specifically for course
options at this level within Universities and
Polytechnics, the series will be of great value to
specialists and research workers in other fields who require
a knowledge of the essentials of a subject.

Other titles in the series:

Experimentation in Biology	Ridgman
Methods in Experimental Biology	Ralph
Visceral Muscle	Huddart and Hunt
Biological Membranes	Harrison and Lunt
Comparative Immunobiology	Manning and Turner
Water and Plants	Meidner and Sheriff
An Introduction to Biological Rhythms	Saunders

TERTIARY LEVEL BIOLOGY

Biology
of Nematodes

NEIL A. CROLL, Ph.D.

Director of the Institute of Parasitology
McGill University, Canada

and

BERNARD E. MATTHEWS, Ph.D.

Department of Applied Zoology
University College of
North Wales, Bangor

A HALSTED PRESS BOOK

John Wiley and Sons

New York – Toronto

Blackie & Son Limited
Bishopbriggs
Glasgow G64 2NZ

450 Edgware Road
London W2 1EG

Published in the U.S.A. and Canada by
Halsted Press,
a Division of John Wiley and Sons Inc.,
New York

Library of Congress Cataloging in Publication Data
Croll, Neil Argo.
Biology of Nematodes.

(Tertiary level biology)
"A Halsted Press book."
Bibliography: p.
Includes index.
1. Nematoda. I. Matthews, Bernard E.,
joint author. II. Title.
QL391.N4C747   595'.182   76-50552
ISBN 0-470-99028-7

Phototypesetting by Print Origination
Bootle, Merseyside L20 6NS
Printed in Great Britain by
Thomson Litho Ltd., East Kilbride, Scotland

Preface

NEMATODES ARE AMONG THE MOST ABUNDANT ORGANISMS ON THE earth. Historically they were studied by a small group of specialists, whose research was directed at the understanding and control of parasitic species. They were often overlooked in biological curricula.

More recently, however, these simple metazoans have been selected as laboratory models for behavioural, genetic, gerontological, nutritional, cryptobiotic and biochemical investigations. This research is now yielding fundamental contributions to biology.

In *Biology of Nematodes,* we have attempted to meet the need for an introductory book which presents a unified view of the whole range of nematodes. Examples have been selected from free-living plant-parasitic and animal-parasitic nematodes, and emphasis has been placed on the similarities rather than on the differences between groups. Wherever possible, line diagrams and photographs have been used to illustrate points in the text.

We are most grateful to the following for permission to use photographs: Drs. D.S. Bertram, E.U. Canning, C.C. Doncaster, A.A.F. Evans, C.D. Green, R.P. Harpur, C. Högger, W.D. Hope, J. Klingler, I.L. Riding, F.W. Sayre, M.K. Seymour, S. Ward and J.M. Webster. We particularly thank our colleague James M. Smith, who took many of the original photographs. We thank the Trustees of the Wellcome Museum, London, for figure 10.1, and Doreen A. Croll and Julie I. Dalziel for proof-reading and their comments on the manuscript.

NEIL A. CROLL
BERNARD E. MATTHEWS

Contents

Chapter 1. THE ZOOLOGY OF NEMATODES   1

Nematodes as worms. The position of nematodes amongst the invertebrates. A zoological description of the nematodes. The classification of the nematodes.

Chapter 2. NEMATODES AS MODELS AND MODEL NEMATODES   12

Developmental biology. Genetics. Gerontology. Toxicity. Biochemistry and physiology. Parasitic models.

Chapter 3. NERVES, MUSCLES AND SENSE ORGANS OF NEMATODES   31

Turgor pressure. Cuticle. Somatic musculature. Nematode neuroanatomy. Neuromuscular physiology of nematodes. The movement of nematodes. Nematode sense organs. Amphids. Phasmids. Cephalic papillae. Caudal papillae. Somatic setae. Ocelli or eyespots. Spicules. Cuticular receptor-effectors. Other sensory cells. The problem of symmetry.

Chapter 4. THE BEHAVIOUR OF NEMATODES   54

The basis of activation. Temporal and spatial patterns of movement. Movement in the egg and hatching. Feeding, Penetration, Air swallowing, Defaecation, Nictating and swarming. Orientation responses. Chemosensitivity and responses. Photosensitivity and responses. Thermosensitivity and responses. Mechanosensitivity and responses. Galvanosensitivity and responses. Geosensitivity and responses, General statement on nematode behaviour.

Chapter 5. THE FEEDING AND NUTRITION OF NEMATODES   79

Digestion and absorption of nutrients. Essential foods. Blood feeding by hookworms and other nematodes. The feeding of whipworms. The feeding of plant-parasitic nematodes. Bacterial feeding. Energetics and nutrition.

Chapter 6. DEVELOPMENT OF NEMATODES   98

Nematode eggs. Embryology. Hatching. Moulting and exsheathment.

vii

Chapter 7.   PATTERNS OF NEMATODE LIFE-CYCLE   120

The nematode parasites of man. The nematode parasites of
vertebrates.

Chapter 8.   THE BIOLOGY OF PARASITIC NEMATODES   135

Seasonality and periodicity. Pathology of nematode infections.
The concept of disease. Site specificity. Pathology. Resistance.
Acquired immunity.

Chapter 9.   SURVIVAL OF NEMATODES   152

Chapter 10.   TREATMENT   AND   PREVENTION   OF
              NEMATODE DISEASES   166

Chemotherapy. Vaccination against nematodes. Physical
methods of control. Control of disease vectors. Biological
control of nematodes. Integrated control of nematodes.

FURTHER READING   185

INDEX   195

CHAPTER ONE

THE ZOOLOGY OF NEMATODES

THE ORIGIN OF NEMATOLOGY IS LOST IN HISTORY, OR EVEN PREHISTORY.
Scholars have interpreted Moses' fiery serpent (Numbers 21, 6-9) as being
the human parasite *Dracunculus medinensis,* and the ancients certainly
knew about parasitic worms. However, as in almost all branches of science,
advancement has been greatest within living memory.

Nematodes were virtually unstudied until the discovery of the light
microscope and its wide availability to biologists; but in the late eighteenth
and nineteenth century the realization grew that they were probably the
most abundant and widely dispersed group of metazoans. It was in the last
century that zoologists concentrated on recording and classifying the
members of the phylum, in their overall attempt to complete what is today
called "classical zoology", a task still far from complete.

The life cycle of the cosmopolitan parasite *Trichinella spiralis* was
worked out between 1822, when the encysted larvae were first seen in
muscle, and 1860, when Leukart published a monograph on trichinosis. In
the same year, Zenker demonstrated that man may acquire the disease by
ingesting infected raw pork. This was a discovery of major importance, as it
was the first time that a generalized disease of man had been shown to be
due to a micro-parasite. It helped to lend credence to the ideas of the
"germ" theory of disease that were then being formulated.

In 1878 Sir Patrick Manson found that blood-sucking mosquitoes in
China transmitted the infective stage of the human parasite *Wuchereria
bancrofti* which causes elephantiasis. This classical discovery led to the
wide recognition in parasitology that invertebrates might be necessary for
the successful transmission of such parasites. It was in the wake of this
discovery that the mosquito was shown to carry malaria, and soon
afterwards the life cycles of the agents of sleeping sickness, plague, and
schistosomiasis were unravelled. This was the phase of "life cycles". These
studies continued into the first half of the twentieth century, and with them

grew the medical, agricultural and veterinary practices which today minimize parasitic infestation.

Plant-parasitic nematodes were publicized by the American nematologist Nathan Augustus Cobb (1859-1932), who was employed by the United States Department of Agriculture to spread an awareness of these plant pests. He described many species, from the Antarctic to the Tropics, and was responsible for initiating the formal study of these nematodes. Cobb's great disservice was that he coined the term "nematology". While focusing attention on the plant pests, he created a zoologically artificial split by distinguishing "nematology", the study of plant-parasitic and soil nematodes, from "helminthology", the study of nematodes parasitic in man and other animals. The latter term is very confusing, because it encompasses the flukes and tapeworms, which do not even belong to the same phylum. This schism in the study of a single zoological entity has obstructed progress and has led to separate Government Departments, research institutes, professional societies and scientific journals.

By the 1930s and 1940s there was an awareness of the medical and economic impact of nematodes on human affairs, so that research increased greatly. Much of the work was done in response to humanitarian and commercial pressures, but zoological syntheses were lacking and, although knowledge grew, much of it was fragmentary and perhaps not sufficiently fundamental. A change in emphasis was marked by the publication of Chitwood and Chitwood's *An Introduction to Nematology* (1950) which brought together a considerable amount of the zoological knowledge about the phylum and its relations. In 1963 Wallace's *Biology of Plant Parasitic Nematodes* was published, which established a major shift in the emphasis of approach for those who study these species. It was followed two years later by Lee's *Physiology of Nematodes* (1965), which gave the first comprehensive account of the physiology of the whole zoological group. This publication was a strong stimulus to the use of experimental techniques on nematodes, and its success may be measured by the numerous physiological studies carried out since.

The small size of most nematodes had limited the possible advances into many aspects of structure and function, until the invention of the electron microscope. In 1971 Bird wrote:

At the moment we are in the midst of a second period of anatomical discovery which started with the development of the electron microscope on a commercial scale in the 1950s, and which is at present gathering impetus...

Through their research, the Chitwoods, Wallace, Lee, Bird and Crofton have succeeded in creating the current increasing biological stature of the study of nematodes. Applied nematology and helminthology can now grow from a thorough basic scientific understanding of the biology of nematodes.

The major biological area undergoing a growth phase is now too wide to be parochially described as "nematological". Biology has moved from studying specific groups of animals to studying biological phenomena that extend across all living organisms. In doing so, it has restructured itself into new biological subdisciplines such as biophysics, molecular biology, genetics, nutritional physiology, gerontology, cryobiology. To investigate such aspects of life, "biological systems" are required, which in themselves may be of limited interest, but which provide a fountain of basic knowledge. We need only consider the contribution of the green alga *Chlorella* to our knowledge of photosynthesis, or the unravelling of the genetic DNA replication system using the bacterium *Escherichia coli,* to see the success of such an approach. The biological systems or "models" are selected on criteria of practical expedience and manipulatability. They must be able to be readily cultured in the laboratory under defined and controlled conditions, be basically simple in organization, and be capable of being handled by available techniques and cultured in quantities suitable for biochemical analysis. Some nematodes described later are being used increasingly as model metazoan systems for broad-based studies into fundamental biology.

Nematodes as worms

Nematodes were described by Aristotle (384-322 BC), and Celsus (53BC-AD7) distinguished between flatworms and roundworms almost 2000 years ago. The Swedish biologist Carl von Linnaeus, who gave biology its first real scheme for classifying animals (*Systema Naturae*), about two centuries ago, placed the nematodes in the class Vermes or worms. Nematodes are indeed "worms", but there are many sorts of animals which are long, thin, flexible and lacking appendages, which are called "worms". Nematode worms are also called "roundworms" or "eelworms". In their ranks they include "pinworms", "threadworms", "hookworms" and "trichina worms". Earthworms and lug-worms, however, are in the phylum Annelida, and the flatworms such as flukes and tapeworms are in the phylum Platyhelminthes; "ringworm" is a fungus, and the balanoglossid or "acorn worms" share a phylum with man, both being in the Chordata.

"Worm" is in fact a greatly abused term, and is of very limited usefulness, as it describes no clear zoological grouping.

The position of nematodes amongst the invertebrates

There is an array of about ten lower invertebrate groups which have been shuffled about and organized into various schemes over the last hundred years. Alliances have been established amongst these and they have been rearranged by subsequent workers.

Classification is never final: it is a continuing and dynamic synthesis of all biological information. It is always important to realize that systematic

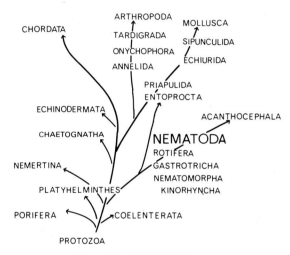

Figure 1.1. The systematic position of the NEMATODA.

Table 1.1. Estimates of the number of species in acoelomate Metazoa

Group	Marine	Soil and Freshwater	Parasitic	Total
Porifera	4,350	500	very few?	4,850
Coelenterata	10,000	17	1	10,000
Ctenophora	90	none	none	90
Turbellaria	many	many	few	1,500
Nemertea	700	few	very few?	750
Rotifera	50	1,680	very few?	1,800
Gastrotricha	210	250	very few?	500
NEMATODA	6,500	8,000	15,000	29,500
Kinorhyncha	50	none	none	50
Priapulida	6	none	none	6
Entoprocta	58	2	none	60
Acanthocephala	none	none	300	300

schemes are attempts to guess at natural relations and need not be correct. The debates that continue between the best informed experts are evidence of the vulnerability of systematics. This is particularly true with the nematodes. Nematodes are related to Rotifera, Gastrotricha, Kinorhyncha, Nematomorpha, Acanthocephala, Priapulida, Entoprocta (figure 1.1, Table 1.1). The separation of the Platyhelminthes from these groups is based on the presence of a pseudocoelom in the Aschelminthes; this is absent in the Platyhelminthes. The Acanthocephala—a small and totally parasitic group—is usually separated into a phylum of its own, as is the Entoprocta.

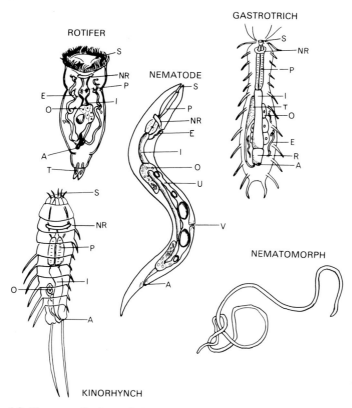

Figure 1.2. The generalized morphology of the Rotifers, Nematodes, Gastrotrichs, Kinorhynchs and Nematomorphs (or Gordian worms). The internal anatomy of the Nematomorph is omitted for clarity. A female adult is drawn in each example, except for the Gastrotrich, many of which are hermaphrodite. (A, anus; E, excretory system; I, intestine; NR, nerve ring; O, ovary; P, pharynx; R, rectum; S, stoma; T, testis; U, uterus; V, vulva).

The views of two authorities on the subject may be usefully quoted:

The taxonomic disposition of the groups Rotifera, Gastrotricha, Kinorhyncha, Nematoda and Nematomorpha has always been troublesome to zoologists (Hyman, 1951).

Reservedly, she placed them all in the phylum Aschelminthes (figure 1.2). Remane (1963) wrote:

The relationships of the so-called Pseudocoelomata is one of the most difficult problems in systematic zoology.

He argued that the "pseudocoelom" may have been developed independently in each group.

No phyletic connection of the classes (or phyla) of Pseudocoelomata is so well founded as to be definitely established... Treatment of the rotifers, gastrotrichs, nematodes, etc., as separate phyla is no solution to the problem, but a failure to face it squarely (Remane, 1963).

Since these writers gave their views, the study of nematodes has developed dramatically. Not only are there now two scientific journals dealing exclusively with researches on nematodes, but at least ten textbooks on the subject have appeared recently. A similar increase in detailed knowledge of the other pseudocoelomate animals is not yet available to allow detailed comparison. It is the intention of this book to follow the consensus of opinion, and to consider the NEMATODA as a phylum.

A zoological description of the nematodes

Nematodes are pseudocoelomate bilateralia with flexible living cuticles and somatic longitudinal muscles for movement. The body tissues are believed to be permanently under pressure. Usually cylindrical, and circular in cross-section, they have no appendages, but may have cuticular bristles or sensory setae. The nervous system has sense organs of separate modalities, central ganglia, and neuromuscular junctions in which the muscle sends a connection to the nerve. There is a stomodaeum and proctodaem lined with cuticle, which are connected by a tubular gut. The gut shows many adaptations to counteract the high internal pressure. The excretory or osmoregulatory system is not protonephridial (no flame cells) and there is no differentiated respiratory or circulatory system. Earlier definitions of the phylum would have stated that there was no metamerism; but this may occur in some nervous systems, although complete segmentation, as in annelids or tapeworms, never occurs in nematodes. Nematodes were earlier thought not to possess cilia but, with the advent of the electron microscope, modified sensory cilia have now been seen in most sense organs; motile cilia have not been seen. The sexes are typically separate, and eggs are laid which pass through four "larval" stages before the adult stage is reached. The larvae are not always morphologically distinct and, because of the incomplete metamorphoses, the term "larva" is really inappropriate. "Juvenile" is the term favoured by Hyman (1951), but "larva" is in such wide usage among nematologists that the term will be retained here. The generalized morphology of adult nematodes is shown in figure 1.3.

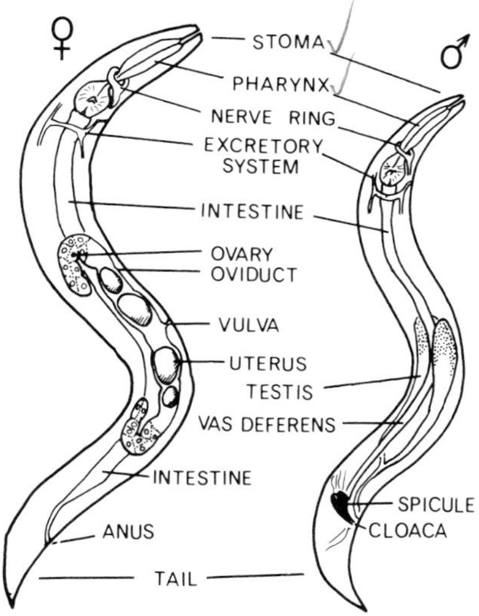

Figure 1.3. The generalized anatomy of an adult male and female nematode

Nematodes may be parasitic in many plants, and in all classes of vertebrates, and most invertebrates including annelids, arthropods, and molluscs. They are abundant in soil and fresh water, and have been recorded from polar regions, mountains and hot springs. Although there are no planktonic nematodes, they abound in the sand and mud of the seas, and have been dredged from great depths from the ocean floor.

Most nematodes are long and cylindrical, but some gravid females tend to become spherical, e.g. *Heterodera* and *Tetrameres*. The smallest adult species which is free-living is about 250μm long, and most species of soil, marine, and freshwater forms are about 1mm long as adults. Parasites of animals tend to be 1-10mm, but many are bigger. *Ascaris lumbricoides* is about 20 cm long and *Dracunculus medinensis* and *Dioctophyme renale* are up to one metre long. The biggest known nematode species is *Placentonema gigantissima*, found in the placenta of sperm whales, which is 9 m (30 ft) long!

Table 1.2. Classification of the phylum Nematoda

Class	Order	Suborder	Superfamily	Selected Genera	Hosts or Habitats
Secernentea (Phasmidia)	Rhabditida	Rhabditina	Rhabdiasoidea	*Caenorhabditis, Panagrellus, Rhabditis, Turbatrix*	free-living: soil, bacterial feeders
			Rhabditoidea	*Rhabdias, Strongyloides*	parasitic: amphibians, reptiles, mammals
	Strongylida	Strongylina	Syngamoidea	*Syngamus*	parasitic: birds, mammals
			Ancylostomatoidea	*Ancylostoma, Necator, Bunostomum*	parasitic: mammals
			Strongyloidea	*Strongylus, Trichonema*	parasitic: mammals
		Trichostrongylina	Trichostrongyloidea	*Dictyocaulus, Haemonchus, Nematodirus, Trichostrongylus*	parasitic: birds, mammals
			Heligmosomatoidea	*Heligmosomum, Nippostrongylus*	parasitic: mammals
		Metastrongylina	Metastrongyloidea	*Metastrongylus*	parasitic: mammals
			Protostrongyloidea	*Protostrongylus*	parasitic: mammals
			Pseudalioidea	*Pseudalius*	parasitic: mammals
	Ascarida	Ascaridina	Ascaridoidea	*Ascaris, Parascaris, Toxocara, Toxascaris*	parasitic: mammals
			Heterocheiloidea	*Anisakis, Porrocaecum*	parasitic: fish, birds, mammals

			Genera	Hosts
	Heterakina	Heterakoidea	*Ascaridia, Heterakis*	parasitic: birds, mammals
		Subuluroidea	*Subulura*	parasitic: birds, mammals
		Aspidoderoidea	*Aspidodera*	parasitic: birds, marsupials
	Oxyurina	Oxyuroidea	*Enterobius, Hammerschmidtiella, Leidynema*	parasitic: insects, amphibians, reptiles, mammals
		Syphacioidea	*Aspiculuris, Syphacea*	parasitic: mammals
Spirurida	Spirurina	Spiruroidea	*Spirura, Tetrameres*	parasitic: birds, mammals
		Thelazioidea	*Gongylonema, Thelazia*	parasitic: birds, mammals
		Gnathostomatoidea	*Gnathostoma*	parasitic: fish, reptiles, mammals
		Acuarioidea	*Acuaria*	parasitic: birds
		Physalopteroidea	*Physaloptera, Proleptus*	parasitic: fish, reptiles, birds, mammals
	Camallanina	Camallanoidea	*Camallanus*	parasitic: fish
		Dracunculoidea	*Dracunculus*	parasitic: mammals
	Filariina	Filarioidea	*Filaria*	parasitic: mammals
		Onchocercoidea	*Breinlia, Brugia, Litomosoides, Onchocerca, Wuchereria*	parasitic: mammals

Class	Order	Suborder	Superfamily	Genera	
	Tylenchida	Tylenchina	Tylenchoidea	*Ditylenchus, Heterodera, Meloidogyne*	parasitic: fungi, plants, insects
		Aphelenchina	Aphelenchoidea	*Aphelenchus, Aphelenchoides*	parasitic: fungi, plants, insects
Adenophorea (Aphasmidia)	Trichinellida	Trichinellina	Trichinelloidea	*Trichinella*	parasitic: mammals
			Trichuroidea	*Capillaria, Trichuris*	parasitic: birds, mammals
	Dioctophymatida	Dioctophymatina	Dioctophymatoidea	*Dioctophyme*	parasitic: mammals
			Eustrongyloidea	*Eustrongylides*	parasitic: birds
	Enoplida	Enoplina	Enoploidea	*Enoplus*	free-living: soil, marine
			Tripyloidea	*Tripius*	parasitic: insects
	Dorylaimida	Dorylaimina	Dorylaimoidea	*Longidorus, Mononchus, Xiphinema*	free-living, plant-parasitic, predatory
			Mermithoidea	*Mermis, Reesimermis, Tetradonema*	parasitic as larvae in insects
	Chromadorida	Chromadorina	Chromadoroidea	*Chromadorina*	free-living: freshwater, soil, marine
			Desmodoroidea	*Desmodera*	free-living
	Monohysterida	Monohysterina	Monohysteroidea	*Monohystera*	free-living: freshwater, marine
			Plectoidea	*Plectus*	free-living: freshwater, marine

The classification of the nematodes

As with their wider zoological relationships, the systematics within the Nematoda must be based on the living species, as no fossil evidence is available. In addition to morphological data, ecological situations can provide strong supporting confirmation of relationships and origins. This is notably the case for the broader groupings such as marine, freshwater and parasitic. Within parasitic nematodes, the form of the life cycles and the systematics of the host species have all provided valuable bases for nematode systematics. Tylenchoidea feed on plants and insects; Ascaridoidea are all parasites of vertebrates; the hookworms, Ancylostomatoidea, are all parasites of mammals, and the Pseudaliidae are parasites of marine mammals. It is very clear that parasitism has arisen more than once in the Nematodes.

It is useful to consider the phylum Nematoda as being divided into two classes: the Secernentea (or Phasmidia) and the Adenophorea (or Aphasmidia). The major orders, suborders, and superfamilies are listed in Table 1.2. This scheme presents a taxonomic organization of the group, and it includes the positions of many of the genera discussed in the subsequent pages of this book.

CHAPTER TWO

NEMATODES AS MODELS
AND MODEL NEMATODES

NEMATODES OCCUR IN A RANGE OF HABITATS WHICH IS UNSURPASSED by any other metazoan group. In spite of their global occurrence and the variety of places in which they can be found, many species are small and difficult to handle in the laboratory. Many of them are parasitic in animals such as man and his domestic animals, and the expense or impracticality of maintaining them is often overwhelming. In the case of parasites, attempts have been made to establish medically and economically significant species in familiar laboratory hosts, or to find closely related species which infect hosts more readily kept under laboratory conditions.

The locust or cockroach is never far away from the laboratory of an insect physiologist; nor is *Drosophila melanogaster* from that of a geneticist. Similarly a growing number of nematodes is encountered as laboratory models.

Some of these models are employed to study basic nematology, but others have a much wider application. Certain species have sufficient features in their size, organization, cellular simplicity and longevity, for them to have been utilized as basic biological models which are now available to many subdisciplines in biology. These models are grown on bacteria or fungi, but they may be cultured "axenically", i.e. without any other living organisms. The culture media have been largely analysed, but in no case are all the essential ingredients of any medium totally defined.

Developmental biology

The emphasis in contemporary molecular biology is moving from unicellular prokaryotes to include multicellular eukaryotes. In this shift, the molecular and genetic bases for development and differentiation are widely sought, but are still elusive. In a recent review, celebrating the 21st anniversary of Watson and Crick's elucidation of the DNA double helix, Brenner (1974) wrote:

12

How do genes specify the complex structures that we find in higher organisms? What controls the temporal sequence of development, the geographical layout of cells and their connections, and specific biochemical differentiation?

Brenner and his colleagues have selected the nematode *Caenorhabditis elegans* (figure 2.1) as a model in which to investigate these questions.

C. elegans is a protandrous hermaphrodite with a generation time of 4-5 days at 20°C. It may be easily grown in the laboratory on bacteria, or axenically in sterile nutrient media. Behavioural mutants can be created by treatment in ethylmethane sulphonate, and behavioural abnormalities selected in the first filial generation. As the species is hermaphrodite, mutant populations may be easily cloned from single adults. The model is further convenient because the larvae may be stored on liquid nitrogen at −196°C. This permits banks of genetic mutants to be stored indefinitely. *C. elegans* has about 2500 genes (Brenner, 1974) and less than 300 neurons (Ward, 1973).

Ward (1973), in a very comprehensive study, examined the orientation of *C. elegans* wild type and of some behavioural mutants to certain chemicals. Comparison of the behaviour of mutants and wild types suggested the location on the body of the receptors mediating the responses of the nematodes. His studies also indicated the nature of the sampling mechanisms used and the form of the locomotory responses of *C. elegans* up attractive gradients.

The nervous system is amongst the most complex of all tissues, as the axons develop along predetermined morphogenetic routes which may run for long distances. It is the hope of these workers to relate the behavioural abnormalities to the ultrastructure of the defective or altered neuromuscular system and ultimately, therefore, to the gene that controls it.

Another type of attempt to examine the basis of development, and therefore gene expression, through the life cycle is being pioneered in Canada (Samoiloff *et al,* 1973). In these experiments, a microlaser beam is being used to cauterize different nerve centres of *Panagrellus redivivus* and *P. silusiae,* and then to follow the animal's subsequent development. Such techniques, though still being developed, provide powerful tools for approaching these basic problems.

Genetics

C. elegans is also being used as a genetic model. Beguet and Brun (1972) and their colleagues in France chose this species to represent the invertebrates.

Homogeneity of the genotype is a critical prerequisite for population

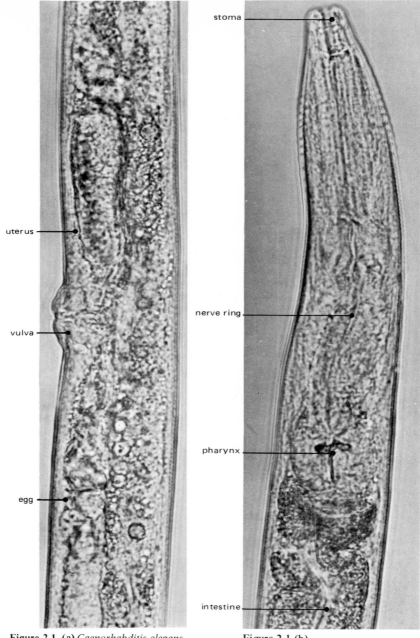

Figure 2.1. (a) *Caenorhabditis elegans.*

Figure 2.1 (b)

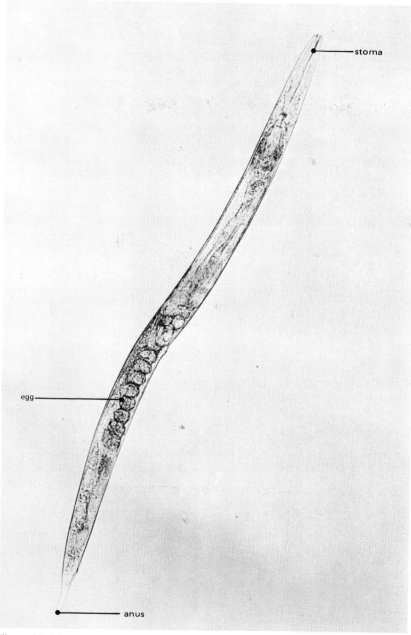

Figure 2.1 (c)

studies of genetic inheritance. The heterogeneity of the gonochoric insect species *Drosophila melanogaster,* largely used as a genetic model, is much greater than that in an alternative, the parthenogenetic rotifer model *Philodina citrina.* The value of an autogamous hermaphrodite such as *C. elegans* is greatly enhanced by its autoreproduction and the homogeneity of its genome.

Various workers have examined the genetic effect of adult senescence on the biological performance of succeeding generations. For the reasons outlined above, *C. elegans* is well suited to such a study. The life cycle of populations feeding on bacteria may be seen to last for about 5 days (i.e. from 4½ to 9½ days). By repeatedly selecting the progeny of "young adults" and "old adults" in successive generations, and counting their offspring, the genetics of senescent fecundity can be studied (Beguet, 1972). Despite certain technical difficulties, the overall result in figure 2.2 was obtained. The fecundity of "old" adults was always below that of "young" adults. The "old" series produced significantly less eggs in the first four generations; then, in the fifth generation and subsequent generations until the ninth, it returned to the previous level of "young" adults. In this experiment, a regulation or genetic feedback for the fecundity of senescence was established. This example is typical of the kind of genetic work possible using *C. elegans.*

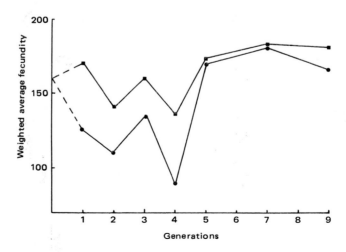

Figure 2.2. General analysis of the fecundity of all hermaphrodites in the young ■——————■ and old ●——————● series of *Caenorhabditis elegans* during the first 9 generations (redrawn from Beguet, 1972).

Gerontology

Ageing is a basic biological property of all multicellular organisms, and probably of all living things; its study, gerontology, is being actively pursued all over the world. Ageing and death, together with an inherent mechanism of variation and mutation in gamete formation, are basic strategies to restrict the size of a population and permit evolutionary change. There are certain strict criteria which must be satisfied in the selection of suitable gerontological models, and nematodes include in their ranks just such appropriate forms. The favoured species are *Caenorhabditis elegans, C.briggsae, Panagrellus redivivus, P.silusiae*, and *Turbatrix aceti*, the vinegar eelworm (Table 2.1). An act of faith has been made by all those who use these models, and it is a parallel one to that successfully made by those who established the structure of DNA. It is that the basic ageing process at its molecular or genetic level is so fundamental that it contains the same elements throughout all biological systems. The results with nematodes are believed to be applicable to man, and to other difficult experimental organisms.

All the above nematodes may be grown in the total absence of other living organisms (an essential for biochemical analyses), on largely defined and rigorously controlled media. This permits genetic standardization, ease of extraction, availability of large numbers, and constancy of nutritional and physiological status. Their maintenance in artificial media further permits the use of additional enzymes, hormones, vitamins and other physico-chemical manipulations. The models must have a rapid generation time and a relatively short life span, producing large numbers of individuals which age quickly and can be analysed by conventional biochemical methods.

For greater simplicity, a gerontological model should be devoid of the capacity to repair cells and should have no cell systems which are being continually renewed. The cell constancy or "eutely", claimed for nematodes, makes them ideal. Some tissues, notably the nervous system and pharynx, do consist of a constant cell number which does not appear to vary between individuals of a species. In tissues such as somatic muscle and intestine, there is an increase in the number of cells during postembryonic development, but no such increase or repair system is known in the adults.

The model must be relatively simple, enabling detailed cytological studies, but it must be complex enough to be fully differentiated into separate organ systems and to show a spectrum of physiological characters. These characters are sometimes very critical features, because ageing must be monitored in some way. Longevity appears to be the simplest parame-

Table 2.1. Some laboratory model nematodes

Nematode species	Generation time (18–22°) days	Estimated longevity (LD50, 18–22°C), days	Longevity in axenic culture (LD50, 18–22°C), days	Estimated peak densities in culture, ml⁻³	Grown on
Caenorhabditis elegans	4–5	10–12	34	200,000	bacteria
C.briggsae	4–5	10–12	34	200,000	bacteria
Panagrellus redivivus	3–4	20–30	30	800,000	bacteria
P.silusiae	3–4	20–30	30	800,000	bacteria
Turbatrix aceti	8–10	50–60	25	150,000	bacteria (in vinegar)
Aphelenchus avenae	7–8	25–30	30–40 (?)	600,000	fungi

ter, and the model should also show other recognizable and measureable features of senescence. Such criteria are being found, and include a decrease in motility (figure 2.3), an increase in specific gravity, the accumulation of "age pigment" and a gradual disorganization of nerve and muscle cells and mitochondria (Gershon, 1970; Zuckerman *et al.*, 1971, 1972, 1973; Epstein *et al.*, 1972).

One of the technical problems in studies on ageing is to synchronize the age of cultures which may contain many millions of individuals. Egg-laying may continue for 4 to 24 days, and therefore after a few generations the population structure of the culture will be very complex and will include all stages of development. Gershon (1970) treated populations of *T. aceti* with the DNA inhibitors, fluorodeoxyuridine and hydroxyurea, and claimed to have reduced DNA synthesis to 10% of the control level. In this way he synchronized the adult age, and he furthermore claimed that growth was normal. The action of these powerful inhibitors is still uncertain for, when the experiments were repeated using *T. aceti* and *C. briggsae*, other workers observed a significant inhibition of growth, blocked maturation and a reduced longevity (Kisiel *et al.*, 1972).

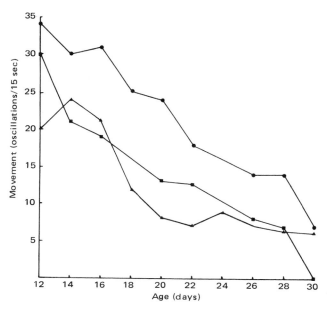

Figure 2.3. The effect of age on the movement of *Caenorhabditis briggsae* (redrawn from Zuckerman *et al.*, 1971).

Toxicity

Various species have been used as "toxicity models" during the development and formulation of nematicides and anthelminthics. In 1952, Peters decided that testing a range of chemical products against the potato cyst nematode *Heterodera rostochiensis* was expensive and lengthy, usually leading to negative results with no new control method. He advocated the use of *Turbatrix aceti,* the vinegar eelworm, as a laboratory model to test new drugs.

Turbatrix aceti was first recorded in Pierre Borel's note *De vermibus aceti* (1656), and was at one time fairly common in commercial vinegar, whence it derived its specific name. The use of oak bark in preparing tan liquors, and of twigs and branches in supporting the growth of microorganisms in vinegar acidifiers, was its probable source in vinegar. In vinegar, *T.aceti* is continuously active, swimming near the surface, feeding on bacteria and completing its life cycle in about 5 weeks.

Because of its great activity and unusual habitat, the usefulness of *T.aceti* is limited, and it has been superseded by *Caenorhabditis* spp. and *Panagrellus* spp. However, recently it has been used to test the effects of drugs in a number of experimental designs. One instance was to measure the growth of populations of *T.aceti* and *C.briggsae* in various doses of the anthelminthic thiabendazole, which is known to be effective against some parasites of vertebrates (figure 2.4).

Piperazine is an anthelminthic used widely against *Ascaris lumbricoides** of man and pigs, and against other nematodes such as the human pinworm *Enterobius vermicularis.* Experiments on nerve-muscle preparations of *A.lumbricoides* have indicated that piperazine acts by inhibiting neuromuscular co-ordination of the nematodes. The pharmacological action of piperazine was tested by Fiakpui (1967), using *C.briggsae. C. briggsae* individuals were observed first in piperazine, then in acetylcholine, and later in both (figure 2.5). From this experiment, it was observed that at high concentrations acetylcholine caused paralysis, probably through causing continuous depolarization leading to a contracted paralysis of the muscles. Piperazine also caused paralysis but, when both acetylcholine and piperazine were applied together, the paralysis was greatly reduced. This led Fiakpui to conclude that piperazine acted as a competitive blocking agent at the post-synaptic receptor site for acetylcholine at the neuromuscular junction.

*There have been various attempts to distinguish between *Ascaris* from man and from pigs, the latter being called *Ascaris lumbricoides* var. *suis* or *A.suum.* In most cases the pig form is studied. *A.lumbricoides* is used throughout in this book.

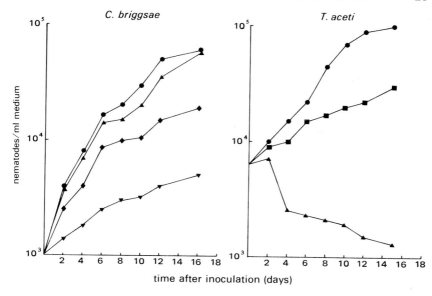

Figure 2.4. Influence of thiabendazole on population growth of *Caenorhabditis briggsae* and *Turbatrix aceti* (redrawn from Vanflateren and Roets, 1972).

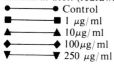
● ────────● Control
■ ────────■ 1 μg/ml
▲ ────────▲ 10μg/ml
◆ ────────◆ 100μg/ml
▼ ────────▼ 250 μg/ml

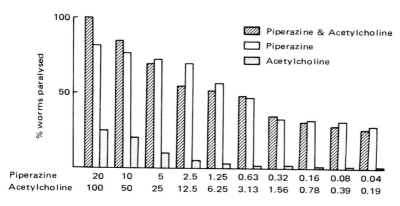

Figure 2.5. Combined action of piperazine and acetylcholine on *Caenorhabditis briggsae* (redrawn from Fiakpui, 1967).

These models and these kinds of techniques have facilitated the growth of knowledge by providing simple, rapid and inexpensive test organisms to try out ideas and to screen compounds which may be used to control nematodes.

Biochemistry and physiology

Numerous studies have been made of the biochemistry and physiology of models to create a background of knowledge about nematodes. Until mass culture techniques were available, these studies were mostly restricted to such atypical forms as *Ascaris lumbricoides* and to other large species which provided sufficient material for biochemical analyses.

Through quantitative analyses and respirometric measurements, researchers have found that nematodes have both a lipid and a carbohydrate energy metabolism, but that different groups of nematodes tend to use one more than the other. Usually animal parasites utilize glycogen and a little lipid while in their free-living stages. In most stages of the plant-parasitic and soil species, lipid metabolism provides the major source of energy. Lipid metabolism, as expected, has been shown to be aerobic. Nevertheless, many soil-inhabiting nematodes are able to survive anoxia or very low oxygen concentrations for extended periods, and to recover after the return

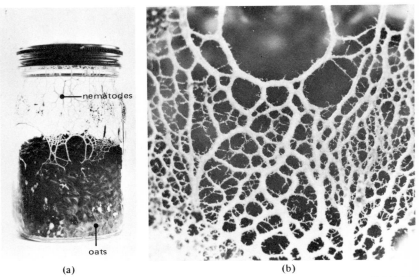

(a)   (b)

Figure 2.6. Kilner jar culture of *Aphelenchus avenae*. Such a culture can yield over 12,000,000 individuals. (material provided by A.A.F. Evans)

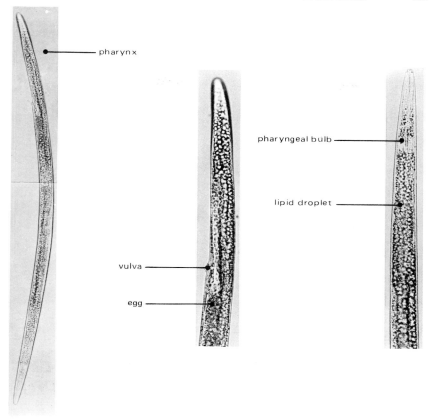

pharynx

pharyngeal bulb

lipid droplet

vulva

egg

Figure 2.7. *Aphelenchus avenae.*

of adequate oxygen. Their survival is due either to tolerance or to the active alterations of their metabolic pathways.

This phenomenon is particularly characteristic of some plant-parasitic nematodes, and a model system has been selected using a stylet-bearing tylenchid, closely related to many agriculturally important species. The model species *Aphelenchus avenae* is a fungal feeder and may be cultured in the laboratory in the absence of bacteria. A cereal such as wheat, oats or barley is autoclaved in jars and inoculated with the fungus *Rhizoctonia solani*. Sterile *A.avenae* are then added and feed on the fungus (Evans, 1970) (figures 2.6 and 2.7). Populations can reach 12,000,000 per jar at 25° C in six weeks; this yields 1.22 g of nematodes, or 0.39 g dry weight.

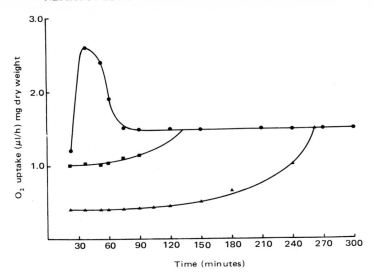

Figure 2.8. Post-anaerobic oxygen uptake for non-feeding *Aphelenchus avenae* and *Caenorhabditis* sp. (redrawn from Cooper and Van Gundy, 1970). ●————● 16 hours without oxygen ■————■ 24 hours without oxygen ▲————▲ 48 hours without oxygen

In a laboratory model situation, *A.avenae* and *Caenorhabditis* sp. have been placed in oxygen-free atmospheres for 16, 24 and 48 hours, and their oxygen utilization measured after the return of oxygen (figure 2.8).

An oxygen debt developed after 16 hours, which resulted in an oxygen consumption above that of the controls. If the nematodes were left longer, i.e. 24 or 48 hours, no oxygen debt developed, and glycogen was used anaerobically, or the nematodes entered a cryptobiotic state. The relationship between the loss of glycogen and lipid in *A.avenae* and *Caenorhabditis* sp. was also examined in aerobic and anaerobic gas phases (figure 2.9). The results demonstrated a switch to glycogen catabolism in the time of oxygen shortage, and a return to lipid catabolism when oxygen became available (Cooper and Van Gundy, 1970).

Parasitic models

The strongest driving force in the study of nematodes has always been the need to control diseases caused by the parasitic forms in organisms such as man, his large domestic animals and his food crops. However, for economic reasons and because of the size of the hosts, these are often the most difficult species on which to experiment. Pressure has therefore developed

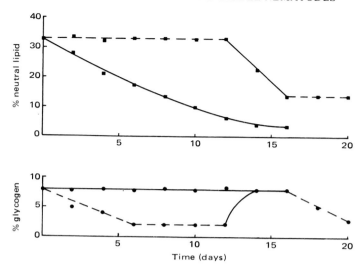

Figure 2.9. Relationship between glycogen metabolism and neutral lipid catabolism in *Aphelenchus avenae* after aerobic and anaerobic exposure (redrawn from Cooper and Van Gundy, 1970). ————— aerobic — — — — — anaerobic

to establish in the laboratory nematodes upon which experiments may readily be conducted. This pressure has been answered in three general ways.

Throughout the 1960s many laboratories tried to grow parasitic nematodes (and other parasitic groups) by *in vitro* culture. Artificial media were designed which would provide sufficient developmental and growth requirements to enable parasites to complete their life cycles without ever entering their hosts. This endeavour has occupied a considerable research effort and has certainly made advances, but the growth and development of parasitic helminths, comparable to that *in vivo* in the host, is not yet a practical proposition. It is reasonable to say that *in vitro* culture has become an end in itself, and has not yet led to the routine production of parasites for experimentation.

Another gambit has been to select strains of the parasites to be studied, which either infect laboratory hosts, or are established by successive passage through laboratory hosts. This approach has certainly yielded successes: for example *Ascaris lumbricoides* can complete its migration in rabbits, as can *Trichinella spiralis* in mice (Table 2.2). In certain programmes it has been necessary to treat some recipient hosts with immunosuppressants in the initial establishment.

Table 2.2. Laboratory models of some parasites of medical and economic importance which are used extensively in research and chemotherapeutic studies.

Nematode	Host	Model Nematode	Host
Ancylostoma duodenale	man	Ancylostoma tubaeforme	cat
Necator americanus	man	Ancylostoma caninum	dog
		Nippostrongylus brasilienis	rat
Strongyloides stercoralis	man	Strongyloides ratti	rat
Haemonchus contortus	sheep	Nematospiroides dubius	mouse
Trichostrongylus spp.	cattle, sheep	Trichostrongylus retortaeformis	rabbit
(Numerous trichostrongyles)		Nippostrongylus brasiliensis	rat
Ascaris lumbricoides	man, pig	Ascaris lumbricoides	rabbit
Enterobius vermicularis	man	Aspiculuris tetraptera	mouse
		Syphacea spp.	mouse
Wuchereria bancrofti	man	Breinlia booliati	rat
Onchocerca volvulus	man	Brugia pahangi	cat, dog
Loa loa	man	Brugia malayi	cat, dog, primate
Other filarials	man	Litomosoides carinii	cotton rat, gerbil
		Dipetalonema witeae	hamster, gerbil
Trichuris trichiura	man		mouse
Trichuris ovis	sheep		
Trichuris bovis	cattle	Trichuris muris	
Trichinella spiralis	man, pigs and others	Trichinella spiralis	mouse, rat

Table 2.2 *(Continued)*

Ditylenchus dipsaci	bulb plants	*Ditylenchus myceliophagus*	fungi
Ditylenchus destructor	potatoes	*Aphelenchus avenae*	fungi
Heterodera spp.	many vegetables	*Aphelenchoides blastophthorus*	fungi
Meloidogyne spp.	many plants	*Aphelenchoides composticola*	fungi
Aphelenchoides spp.	many plants		

Finally efforts have been made to find "model parasitic nematodes". These are species which are closely related taxonomically and biologically to the medically and economically important species, but which parasitize familiar laboratory hosts like mice, rats, hamsters, cats and birds. By studying these model species, basic information has been acquired which has led to the understanding and control of economically important forms. In the laboratory, experiments may be replicated adequately, the genetic variability of hosts and parasites may be better controlled, and variables such as diet, climate and daily or seasonal fluctuations may be eliminated. Table 2.2 summarizes some of the most familiar laboratory models of parasitic nematodes.

Nippostrongylus brasiliensis is probably the most widely studied model for nematodes parasitic in animals. The infective larvae resemble hookworms in that they penetrate through the skin of rats, and then migrate to the intestine where they live in a manner similar to many strongyle and trichostrongyle species of medical and veterinary importance. *N.brasiliensis* has been the subject of many hundreds of researches. These have been concerned with attempts to establish it in culture, to examine its ultrastructure, to understand the mechanism of its rejection by the host's immune response, and to investigate its biochemistry, physiology and behaviour.

If rats are given a major infection of about 1000 infective larvae of *N.brasiliensis,* an immunity develops in the rat and the adult worms are rejected within about three weeks. This has been used as a central "model" reaction of immunity in parasitic nematodes. It has been found that *N.brasiliensis* adults persist for much longer periods if the infective larvae are "trickled" in at the rate of a few per day (Jenkins and Phillipson, 1972). Results such as this have a great impact on our thinking about the nature of immunity in parasitic helminths. A similar procedure has been used to test or "screen" thousands of drugs in the formulation of new effective anthelminthics. *Nematospiroides* (*Heligmosomum*) *dubius*, also in mice, has been used as a readily available model for the trichostrongyles. *Strongyloides ratti*, in rats, has been available as a model species to investigate the human parasite *Strongyloides stercoralis*.

The human pinworm *Enterobius vermicularis* is a widespread human parasite—sometimes called "the scourge of the boarding schools"—and is very common in England. Two closely related species, *Aspiculuris tetraptera* and *Syphacea obvelata*, are very common in mice, and provide good models for experimental work.

Trichinella spiralis, a parasite of pigs and man—infecting man when he consumes undercooked pork—is a species with a very wide host range. As a result, rats are convenient hosts which are used in immunological and chemotherapeutic studies. The human whipworm *Trichuris trichiura*, which infects hundreds of millions of people, is a member of a most successful genus, species of which infect sheep, cows, goats, pigs and many other mammals. *Trichuris muris* is the laboratory model for this parasite, and is used to gain basic information about the genus.

The filarial worms, which infect hundreds of millions of people and many more of their domesticated animals, form one of the most significant groups of nematodes. Filarial worms like *Wuchereria bancrofti*, transmitted by mosquitoes, and *Onchocerca volvulus*, carried by the blackfly *Simulium* spp., have complex life cycles. These and other human parasites do not infect convenient laboratory hosts—a fact which has undoubtedly retarded their study. The most widely used model is *Litomosoides carinii*, which lives in the pleural cavity of the cotton rat *Sigmodon hispidus* and is transmitted by the tropical rat mite *Bdellonyssus bacoti* (figure 2.10). A second model, used particularly in the last decade, is *Dipetalonema witeae*, a filarial nematode living subcutaneously, which was isolated from the Libyan jird *Meriones libycus*. It can be established in the Mongolian jird *Meriones unguiculatus* (the "gerbil") and in the hamster. It is transmitted by the tick *Ornithodorus tartakovskyi*, but is resistant to the antifilarial compound diethylcarbamazine, which is used widely in human filariasis. These models live in different sites from *W.bancrofti*, which lives in the lymphatics, and are transmitted by acarines, not insects. Such wide biological differences have undermined confidence in their use.

With the discovery in the late 1950s of a number of lymphatic-dwelling filarial worms assigned to the genus *Brugia*, life-cycle studies were possible on a species more closely related to *W.bancrofti*. *Brugia malayi* is a human parasite in the Far East. *Brugia pahangi* and *B.malayi* in cats have provided

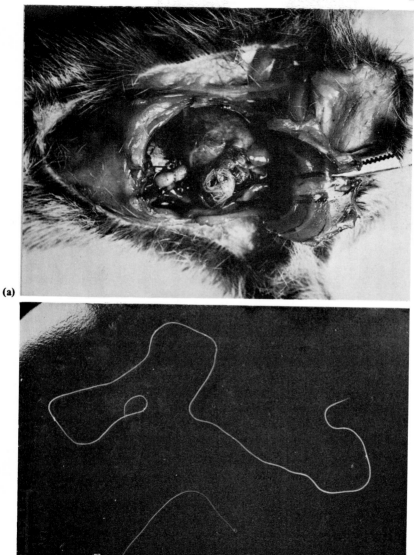

Figure 2.10. (a) Coil of adult worms of *Litomosoides carinii* in the pleural cavity of a cotton rat *Sigmodon hispidus*, as exposed by dissection. (b) Female and smaller male *L. carinii* (Bertram, 1966). Note spiral at hind extremity of male.

a great deal of information, especially about pathogenesis, immunity and the duration of infections. Nevertheless, the prohibitive cost of using these large animals has inevitably hampered the rate of progress.

Workers in South-East Asia have been examining wild mammals, and have discovered species such as *Breinlia sergenti* in the slow loris *Nycticebus coucang,* but have been unsuccessful in transmitting it in the laboratory. *Breinlia booliati,* from the thoracic and peritoneal cavities of the tree rat, *Rattus sabanus* in Malaysia, was described for the first time in 1972. Some success has been recorded in infecting the laboratory albino rat with infections carried by the mosquitoes *Aedes togoi* and *Armigeres subalbatus.* This species could become a standard model for a devastating group of nematode parasites for which no adequate model exists as yet.

Amongst the plant-parasitic nematodes, the problems of "domestication" are smaller but nevertheless similar. Species on bananas, cotton, coffee or tea cannot easily be reared in temperate climates, and observations on all species in soil are most restricted.

Ditylenchus myceliophagus, Aphelenchus avenae, Aphelenchoides blastophthorus and *Aphelenchoides composticola* (Table 2.2) may be mentioned as plant parasites which are easily cultured in the laboratory on fungi, in agar which is free from bacteria. They can yield large quantities of nematode material in a few weeks, and can be observed throughout their life cycles.

Parasitic laboratory models have been discussed at some length near the beginning of this book, because the reasons for using them, and the data that they provide, are at the hub of modern nematology.

CHAPTER THREE

NERVES, MUSCLES AND SENSE ORGANS OF NEMATODES

THERE WILL ALWAYS BE A DEBATE AMONG SPECIALISTS AS TO WHICH single piece of research has most influenced and advanced fundamental nematology. Whichever studies are considered, the classic contribution of Harris and Crofton (1957) would probably be in everyone's short list. Their paper, largely unchallenged after 20 years, provides the basis for understanding the functional organization of nematodes. These authors marvelled at the uniformity of nematodes, a group whose members appear similar and yet frequent such a diversity of habitats.

This uniformity can hardly be ascribed to simplicity of organization. Well below the level of complexity of the nematodes, the protozoa, sponges and coelenterates show a rich variety of adult form and of larval stages within comparatively restricted taxonomic or ecological groupings.

A bridge, a ship, or an aeroplane are recognizable at once because their design is based necessarily and largely on the mechanical forces which play such a predominant part in their economy. Is it in such limitations that we find the key to the underlying uniformity in the nematodes? (Harris and Crofton, 1957, p.116).

The dynamic nematode, as conceived by Harris and Crofton, may be understood through the interaction of hydrostatic turgor pressure, an elastic cuticle and the somatic musculature.

Turgor pressure

When a fluid-filled glass pressure gauge was inserted into a living adult *Ascaris lumbricoides,* the body contents were found to be under an internal turgor pressure. This pressure varied widely from 16 to 125mmHg, and could reach 225mmHg for a few seconds. The variation was not only between individuals, but within a single individual, and followed a rhythmic rise and fall at an interval of about 30 seconds. The mean value for adult *A.lumbricoides* was 70mmHg (95 cm of water). Harris and Crofton (1957) concluded from this and related data that the fluid-filled pseudocoelom was in open communication, and that the changes in pressure caused

Table 3.1. Turgor pressures developed in a range of invertebrates, from various sources (quoted in Harris and Crofton. 1957).

Species	Turgor pressure (mmHg)
Calliactis parasitica (coelenterate)	0–10
Arenicola marina (annelid)	10–20
Lumbricus terrestris (annelid)	1–21
Peripatopsis (onychophore)	2–15
Potamobius fluviatilis	10–18
Carcinus maenas (crustacean)	3–19
ASCARIS LUMBRICOIDES (nematode)	16–125

isometric extension along the body. These internal pressures were of an order of magnitude higher than measured pressures in many other invertebrates (Table 3.1).

The pressure inside *A.lumbricoides* could be directly and artificially altered by ligaturing half an adult onto a water manometer. At pressures of up to 15mmHg the half-worm lay inert; at 45mmHg occasional contractions occurred, and at 60-100mmHg there were regular and sustained bursts of locomotory waves. Contractions became irregular above 150mmHg and ceased at pressures of over 200mmHg.

These experimental findings relate closely to the observation that strips of body wall, with muscles attached, will show spontaneous rhythmic contractions when loaded with weights of 10-20 g, but not with weights of 25-30 g. Assuming a cross-sectional area of about 15mm², a tension of 10-20 g corresponds to an internal pressure of 75-100mmHg. As the mean turgor pressure for *A.lumbricoides* was 70mmHg, this provides a reasonable check on the system proposed and suggests a form of coordination between muscle contraction and turgor pressure.

Regrettably, measurements on *A.lumbricoides* are the only detailed records of turgor pressure in nematodes, and it has been said that *A.lumbricoides* may be to nematodes what the blue whale is to mammals. The clarity and elegance of this work has led to its unqualified acceptance as the *modus operandi* for all nematode work. Active nematodes are indeed always turgid (and dehydration causes immobility) but the extent of a real pseudocoelomic fluid in some of the smaller forms has been questioned and "bits" of chopped nematode can still contract after the apparent release of turgor pressure. In small nematodes the cuticle itself may provide the major component that opposes longitudinal muscle contraction. Implicit in the *A.lumbricoides* proposals is the contribution to the internal pressure made by the other internal organs. It is perhaps the incompressibility of these organs in the smaller forms which provides the necessary turgor.

The arrangement of the feeding pump and of the reproductive and excretory systems with their "pressure valves" at the body orifices may all be interpreted as modifications to counter the high internal turgor pressure. Such modifications occur throughout the phylum Nematoda, both large and small, and provide indirect support for the ubiquity of the Harris and Crofton theory.

Cuticle

Turgor pressure can be maintained only if an increase in volume is resisted by an outer covering. Just as in a balloon, the cuticle of nematodes must be mechanically strong, elastic and capable of some extension. *A.lumbricoides* may increase in length by 10-15% and, under conditions of hypotonic stress, some species can increase in length by over 25%. The complex multilayered nematode cuticle has been extensively studied and is fully reviewed by Bird (1971). In the cuticle of *A.lumbricoides* there is a meshwork of spiral, relatively inelastic fibres. These fibres run counter to one another and permit anisometric extension of the cuticle. They work in the form of a trellis or lattice, and may be best visualized as a series of unit parallelograms (figure 3.1).

In each parallelogram there will be a critical position from which the length and diameter can change without altering the volume. In the diagram, this may be illustrated when Q is $55°$. If the angle Q is larger than $55°$, an increase in length will more than compensate for a decrease in the diameter. Below $55°$, lengthening of the worm decreases its volume. Direct measurements of fixed specimens of *A.lumbricoides* showed a mean angle for Q of $75°30'$. For such an angle, length and volume changes are directly related. For a value of Q greater than $55°$ (i.e. the situation in *A.lumbricoides* and presumably some other nematodes) contractile elements must be longitudinal. Any contraction of the longitudinal muscles tends to shorten the body and reduce the volume. Since the fluid contents are incompressible, the increased pressure is transformed to an increase in width. It is possible that the rhythmic rise and fall in turgor, measured in *A.lumbricoides,* relates to the pressure exerted by the elastic cuticle and other organs, and the additional pressure developed by muscular contrac-

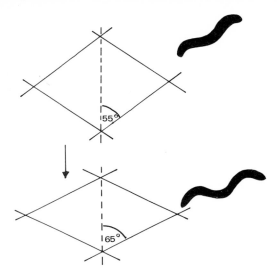

Figure 3.1. The body wall of *Ascaris lumbricoides* and many other species is a complex structure which contains two sets of spiral fibres which run in opposite directions. When the body is shortened and fattened, the angle between the fibres is decreased; but when lengthened, the angle is increased. The range of body shape is restricted by the fibres and the maintenance of the hydrostatic turgor.

tions. The proposed system of interacting pressure, cuticle and muscle is capable of maintaining a static equilibrium over a wide range of body volume. This would permit the ingestion of fluid and its expulsion in defaecation.

Since the proposals for *A.lumbricoides* were published, it has been found that spiral fibres are not essential, and an anisometric cuticle can be constructed of alternate rigid and flexible annulations. This notion has been elaborated for the very small nematodes where the cuticle has been envisaged as a coiled spring because of its striations, annulations and incompressible matrix layer. They may be antagonistic to the muscular effort. These detailed variations in no way compromise the central idea of an antagonism between muscular contraction and the elastic cuticle.

Somatic musculature

Sinusoidal or undulatory propulsion is caused by the asymmetrical co-ordinated contraction and relaxation of somatic muscles down the length of the body. The "typical" nematode, for the purposes of movement, may be taken to be two opposed hemicylinders. The phases of contraction or relaxation of the somatic muscles of one hemicylinder are complemented by the opposed phases in the other hemicylinder. The hemicylinders are not actually "lateral", but are dorsal and ventral in the morbid anatomy of the nematode, and in two dimensions nematodes swim on their sides. The definition of "dorsal" and "ventral" is based on the position of the anus, excretory pore and vulva, which are considered to open on the ventral side.

Four longitudinal strips of muscle cells run along the quadrants of the body, each separated by a hypodermal cord. In sinusoidal movement the dorso-lateral pairs contract together, and these are opposed by the ventro-lateral pairs.

The application of the electron microscope to nematodes has provided a detailed knowledge of nematode muscles (reviewed by Bird, 1971). The contractile element is attached to the hypodermis and cuticle at its distal end; and the proximal end of the spindle-shaped cells, the cell body, hangs freely in the pseudocoelom. The somatic muscles are obliquely striated, and contain thick and thin myofilaments which could be actin and myosin. The myofilaments may slide over one another in a way comparable, but not identical, to the striated muscles of vertebrates. The muscle cell body contains the nucleus and mitochondria, and a projection which contacts the nerve.

Contraction of these muscles is opposed by the turgor pressure, which

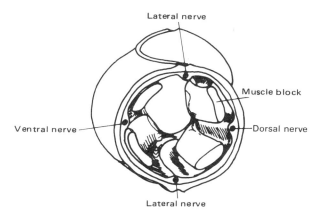

Figure 3.2. Three-dimensional reconstruction of nematode body wall. A section of body shows how dorso-lateral and ventro-lateral muscle blocks contract in opposition to propagate waves in movement.
N.B. In movement the "dorsal" and "ventral" nerves are actually lateral. The ridges along the body are the lateral lines.

also acts to transmit pressure changes, hydraulically, throughout the organs of the nematode. The extent of change in shape, through increasing length and decreasing width or vice versa, is accommodated by the fibrillar layer of the cuticle and the overall elasticity of the cuticular layers.

The contractile part of the muscle cell may be arranged in a number of ways, but it is always attached to the hypodermis and cuticle, and its contraction causes an immediate deformation of the cuticle and a change in the posture of the body.

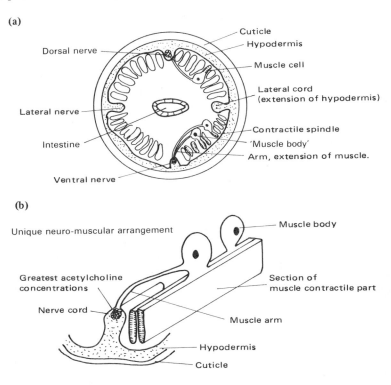

Figure 3.3 (*a*) Transverse section of generalized nematode to show the features of the neuromuscular physiology. The cuticle surrounds the body. Underneath it is the hypodermis with four expansions, one in each quarter of the body. Embedded in these hypodermal cords are the nerves. The muscle cells are arranged between the cords, and each sends a projection to the nerve.

(*b*) A reconstruction to show the organization of two muscle cells and their connection to the nerve. Each cell has an elongated longitudinal contractile portion, a cell body with the nucleus and an arm. The neurotransmitter, acetylcholine, is most concentrated at the nerve-muscle synapse.

The muscle cells are complex and large; each is composed of three distinct parts (figure 3.2). The longitudinal contractile portion is long and attached directly to the hypodermis and cuticle. Above this is a balloon-like "muscle cell body"; this portion is nucleated and lies free in the pseudocoelom. Extending from the cell body and contractile element is the tubular "muscle arm". The arm enters the dorsal or ventral nerve, and makes a synaptic connection with the nerve. While not being unique in animals, an arrangement in which the muscle sends out a connection to the nerve, and not vice versa is most unusual (figure 3.3).

Nematode neuroanatomy

Nematodes neither carry out their behavioural activities at once nor are such activities random. There is a clear requirement for pacemakers or central coordinators in the organization of their behaviour. The correlation between the behaviour and the neuroanatomy of nematodes has yet to be formalized. Nevertheless, the description below is designed to suggest the functional nature of nerve trunks. It is based on small fragments of information from various sources.

The main nerve centres of the central nervous system are the circumpharyngeal nerve commissure ("nerve ring") which encircles the pharynx, and the "anal" or "rectal" ganglia which form a cluster of ganglionic centres near the tail (figure 3.4). Longitudinal nerves which run anteriorly or posteriorly are arranged radially or bilaterally (figure 3.4). The "nerve ring" itself is usually highly asymmetrical, being twisted towards the large retrovesicular ganglion on the ventral side. Four main longitudinal nerves connect the "nerve ring" with the "anal" ganglia, each being embedded in the hypodermis. These are the "dorsal" and "ventral" nerves and the lateral nerves; some species have 4, 8 or even 10 lateral nerves. All these somatic trunk nerves are variously connected by radial commissures.

Electrophysiological evidence strongly suggests that the spontaneous contractions of longitudinal somatic muscles are modulated and coordinated by the ganglionated ventral and dorsal nerves. Action potentials of up to -28 mV have been recorded from these nerves which in *Ascaris* move at about 6 cm/s. The ventral twisting of the nerve ring, the large size of the ventral nerve, the retrovesicular ganglion and the conspicuous ganglia of the ventral nerve, all suggest a dominant role for the ventral nerve in the coordination of movement. Also on comparative grounds, the main motor nerve in rotifers, oligochaete annelids and arthropods, is the ventral nerve. Widely separated muscle cells of the dorsal 'field' in

A.lumbricoides show rhythmic action potentials in phase, while those of the ventral field show another distinct phase. This strongly implicates a separate pacemaker, if not common motorneurones and these probably involve the dorsal and ventral nerves.

If the dorsal and ventral nerves are largely motor, then the lateral nerves are probably sensory. In the acanthocephalans and gastrotrichs, which have a predominantly radial symmetry, there are large lateral nerves; but no neurophysiological studies appear to exist on these groups.

Eight nerves pass forwards from the nerve ring which, unlike the somatic nerves, run in the pseudocoelom. These include six papillary nerves which connect with papillae, and two amphidial nerves which connect with the amphids. A powerful nervous circuitry therefore connects the anterior sense organs, the nerve ring, the longitudinal motor tracks, the somatic muscles and the anal ganglia (figure 3.4).

There are two so-called "sympathetic" nervous systems in nematodes (Bird, 1971): the pharyngeal-enteric system and the rectal-enteric system. These have been described in all species examined. The pharyngeal-enteric sympathetic nervous system is not intimately connected to the central nervous system, although connections are known in some species. There is a considerable body of behavioural evidence to separate the functions of feeding and somatic motility (Croll, 1972). The rectal-enteric sympathetic nervous system does not appear to be as neurally isolated from the central nervous system as the pharyngeal-enteric sympathetic nervous system.

Neuromuscular physiology of nematodes

The neuromuscular pharmacology and electrophysiology of *Ascaris lumbricoides* and one or two other large nematodes have been fairly tho-

Figure 3.4. The generalized nematode nervous system. This figure includes the anterior stylet typical of plant-feeding species:
(a) whole individual,
(b) anterior end,
(c) circumpharyngeal commissure and major associated nerves and ganglia,
(d) section through intentinal area,
(e) posterior end.
Stippled area, alimentary tract; hatched area, somatic muscles.
Abbreviations: ag, amphidial ganglion; amph, amphid; an amphidial nerve; cp, cervical papillae; cpc, circumpharyngeal commissure; dg, dorsal ganglion; dn, dorsal nerve; drg, dorsal rectal ganglion; dvc, dorso-ventral commissure; elg, external-lateral ganglion; int, intestine; imf, intestinal muscle fibres connecting the body wall and the intestine; lc, lateral cord; lg, lumbar ganglion; lm, labial muscles; ln, lateral nerve; pg, papillary ganglion; pn, papillary nerve; rect, rectum; rvg, retrovesicular ganglion; sm, somatic muscle; spm, stylet protractor muscle; sspm, secondary stylet protractor muscle; st, stylet; vg, ventral ganglion; vn, ventral nerve.

Figure 3.4

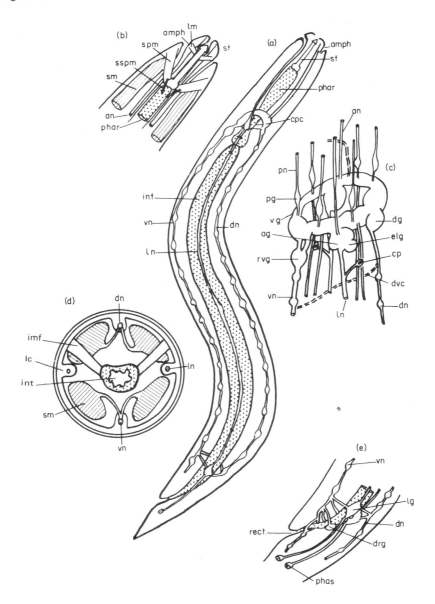

roughly investigated by del Castillo and his collaborators (reviewed, del Castillo and Morales, 1969). The somatic musculature is the only portion of the worms which has been sufficiently studied to permit some general remarks to be made.

It has been found that muscle cells undergo spontaneous rhythmic myogenic depolarizations. As a result, the concept of a *neurocratic* neuromuscular system has developed in which excitatory and inhibitory impulses accelerate or depress the rate of muscular activity. The electrical activity of each somatic hemicylinder, i.e. the dorsal and ventral muscle fields, is integrated and each is probably under common pacemaker control.

The candidates for excitatory and inhibitory transmitters have received some attention. When Lee (1962) stained sections of *A.lumbricoides,* he found, in addition to digestive esterases, that the sense organs, the spicule core, the nervous system and the neuromuscular synapses, were all positive for esterases. This work has now been confirmed on many different species and has been repeated using specific cholinesterase stains such as acetylthiocholine iodide. At the electron-microscope level, McLaren (1972) has mapped the distribution of the cholinesterase-positive material in amphids and other sense organs, and has confirmed its identity with specific cholinergic inhibitors such as physostigmine. Acetylcholine and acetylcholinesterase have been isolated from homogenates of nematode tissues and, although this is not conclusive proof of its involvement in neuromuscular physiology, when considered with the other evidence, it is highly indicative.

Acetylcholine can depolarize somatic muscle cells of *A.lumbricoides* if applied topically to isolated nerve-muscle preparations at the neuromuscular junction. This is also the region of the most dense esterase staining. Indirect evidence of cholinergic nervous transmission has been acquired from experiments with drugs used to control nematode parasites (Chapter 10). The organophosphate and carbamate drugs immobilize nematodes and, by analogy with insecticides such as DDT, are believed to act on nervous coordination. Piperazine hyperpolarizes muscle preparations and antagonizes the depolarizations caused by acetylcholine (figure 2.5). Tubocurarine, neostigmine, succinyl choline chloride and physostigmine are amongst the drugs known to effect nematode movement and are inhibitors of cholinergic transmission.

While it is reasonable to assume that excitatory nervous transmission occurs in the central nervous system of nematodes, this assumption need not apply to the sympathetic nervous systems. Further work is needed because even Vitamin B_1 (thiamine) causes the excitation of muscle strips.

The identity of the inhibitory transmitter, if indeed one exists, is much less certain. On electrophysiological grounds γ amino butyric acid (GABA) has been suggested, but there is no further area of support for this. The classic division between cholinergic and adrenergic sympathetic and parasympathetic nervous systems in vertebrates, is now thought to be a gross oversimplification, and in time it will probably be demonstrated that the picture is also much more complex in nematodes. Recently, for example, a single cell in the brain of the mollusc *Helix aspersa* was found to be both cholinergic and dopaminergic.

Interest is now being taken in the indole alkyl amines or biogenic amines in nematode coordination, a development which has been partly stimulated by their demonstration in parasitic flatworms. Serotonin (5-hydroxy-tryptamine), which is very widely distributed in both vertebrates and invertebrates, decreased the rate of somatic contractions in *Phocanema decipiens*; and Croll (1972) suggested that it may coordinate endogenous activities of nematodes. Norepinephrine (22 ng/g tissue) and epinephrine (2 ng/g of tissue) have been extracted from the tissues of *A. lumbricoides*. Small amounts of epinephrine and considerable amounts of norepinephrine were demonstrated in larval *Haemonchus contortus*. In *H. contortus* larvae, Rogers and Head (1972) found a dramatic increase in norepinephrine following stimulation for development. This result associated biogenic amine with morphogenesis, possibly through neurosecretion.

There is a very effective method of localizing indole alkyl amines in tissues using fluorescence microscopy; Anya (1973) has applied this to the nematode *Aspiculuris tetraptera* and mapped the distribution of 5-HT in muscles, the pharynx and in the male reproductive system. Serotonin also activates the vulval and vaginal muscles in a number of nematodes.

Indolealkyl amines have been found in a number of places and in association with separate activites. The nematode evidence suggests that they will be shown to be very significant and are particularly implicated in endogenous actions which may not be coordinated by the central nervous system. They may also act as inhibitors in neuromuscular coordination.

The movement of nematodes

Nematodes push against their environments and propel themselves by using the forces which act normally to their surface. The usual form of the movements generated by the nematodes is a series of sine waves which move along the body. Most waves pass backwards, and the dimensions of the waves may be influenced by the environment. For nematodes about 1 mm or less long, in bodies of fluid greater than their own diameter, the undulations are very wide and amplitude relatively great. As the meniscus approaches their own diameter, the waves become smaller, and this tendency increases as the depth is further reduced. Backward movements or reversals with consequent forwardly directed waves occur either spontaneously or in response to entry into unfavourable conditions.

Existing studies on locomotion have concentrated on nematodes moving in water, on sterile agar, in thin water films on glass, or between grains of sand. This is analogous to describing human movements by observing only people running around a racetrack. In different situations, the wider versatility of the basic organization becomes more apparent, and nematodes begin to carry out their additional physiological functions. These include: orientation to or from food exudates, sex-attractant and other attractive or noxious chemicals; selection of feeding sites; ingestion of food; intestinal displacement and defaecation; alignment of bodies for insemination and oviposition. The successful completion of all of these activities requires a higher order of co-ordination and behavioural integration than is allowed for in the basic model system of locomotion. Unfortunately the behavioural components of all these activities have not been extensively studied. Detailed analyses of copulation, defaecation and oviposition are, for example, still limited to only two or three descriptions. For a fuller account of the behaviour of nematodes see Chapter 4.

Nematode sense organs

In the last 5–10 years, and mostly in the 1970s, there has been a phenomenal increase in our knowledge of the structure and identity of nematode sensory receptors. These advances have resulted directly from the application of the electron microscope to nematode material. There is no direct electrophysiological evidence available to support the argument that these structures are sensory receptors, i.e. no action potentials have been recorded from sensory neurones upon their stimulation. This having been stated, the indirect evidence is overwhelming.

Nematodes respond to separate sensory modalities, including chemical, mechanical, photic and thermal stimulation. There are less clear reports of orientation to gravity and electrical potentials. The structure, innervation and position of the sensory receptors are consistent with their function as receptors, and often compare closely with similar sensory cells in other animal groups (figure 3.5).

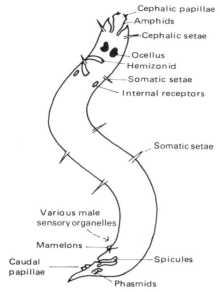

Figure 3.5. Generalized nematode to show sense organs. No species is known to have all these sense organs, but those cells believed to be sensory are included in this figure.

Amphids

Usually just behind the anterior tip of nematodes are paired lateral pits which lead into the amphids. The amphids are typically large spaces filled with modified cilia, known collectively as "sensilla". These cilia may have many or a few microtubules, and have neither the two central microtubules nor the basal body typical of motile cilia. The sensilla are connected via the amphidial nerve to the circumpharyngeal nerve commissure. The external opening of amphids in some marine nematodes may be elaborated into a complex helical spiral, or into a shield-like structure. The functional significance of these structures is obscure.

The amphids are associated with amphidial glands which may be very large. In hookworms and other nematode parasites of the gut, the amphidial glands have been shown to secrete large amounts of protein, including the enzyme acetylcholinesterase. These secretions act as antigens in the immune response of the host, and as such have created considerable interest. In free-living soil and marine nematodes, gelatinous mucus has been found in the amphidial pores.

It has been assumed that amphids are chemoreceptors and, when malformed as in "blister mutants" of *Caenorhabditis elegans,* the response to chemical gradients was lost (Ward, 1973). The role of amphids as secretory glands is also becoming clear, but its significance is unknown.

Phasmids

The phylum Nematoda has been divided into two classes: the Phasmidia and the Aphasmidia, on the presence or absence of phasmids. Phasmids are found at the tail end of nematodes and are paired lateral pits which lead into spaces and may be associated with glands. The arrangement is similar to that of amphids, but at the tail end. Modified cilia have been found in those that have been examined with the electron microscope, and it is postulated that phasmids are probably chemoreceptors. The cilia connect with the rectal ganglia and the lateral nerve via phasmidial nerves (figure 3.5).

Cephalic papillae

A series of 6, 8 or 12 cephalic papillae is arranged around the mouth of nematodes. These are simple structures with modified cilia which do not open onto the surface. In the filarial nematode *Dipetalonema viteae,* in which these papillae have been carefully researched, each papilla is composed of a single large cilium which extends into the cuticle.

Caudal papillae

On the tails of adult males, and usually arranged around the cloaca, there is often a series of papillae which have always been assumed to be sensory and to be involved in copulation. These vary in their ultrastructural details from the cephalic papillae, at least in *D.viteae,* one important difference being that caudal papillae may open through the cuticle to the exterior. These papillae are probably sensitive to mechanical or chemical stimulation.

Somatic setae

The bodies of many free-living soil, freshwater and marine nematodes may have many somatic setae. In *Chromadorina bioculata,* for example, there are four rows of somatic setae, two subdorsal and two subventral. There are 130–132 setae in females and 124–128 in males.

From detailed microscopical studies, Maggenti (1964) concluded that the labial setae and/or papillae of the marine form, *Thoracostoma californicum* are mechanoreceptors. The setae of *T.californicum* are connected via neurones to nerves in the lateral fields. Ultrastructural observations on somatic setae of *Chromadorina bioculata* and *Enoplus communis* have demonstrated that these are cuticular extensions, surrounded by areas of weakness in which they are assumed to rock. Passing up through the hypodermis and into the cuticle are modified cilia, and it is thought that sensory input results from distortion of the cuticular portion of the seta (figures 3.6, 3.7).

Ocelli or eyespots

Amongst the free-living nematodes in marine and freshwater habitats,

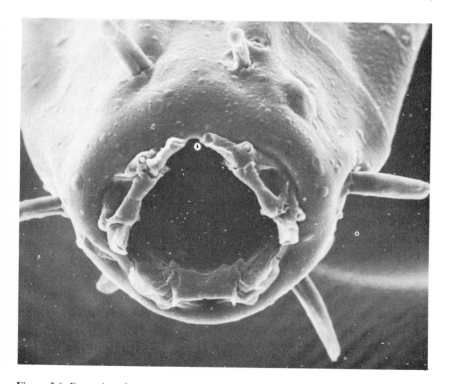

Figure 3.6. Examples of nematode circular projections
(a) stereoscan electron-microscope view of the head of the marine nematode *Enoplus communis*. The ring of six stout projections or setae, are thought to be mechanoreceptors.

Figure 3.6 (b)

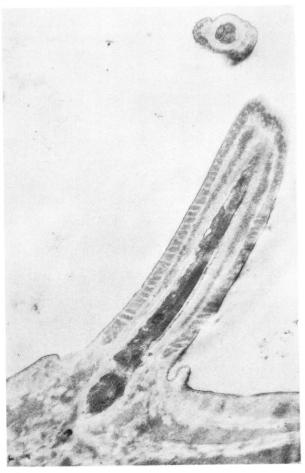

(b) is a longitudinal section taken with a transmission electron microscope through a seta of *E. communis.* Near its base, the seta may be seen to be in a depression in which it can rock; the "sensillum", or modified sensory cilium runs in the middle of the seta.

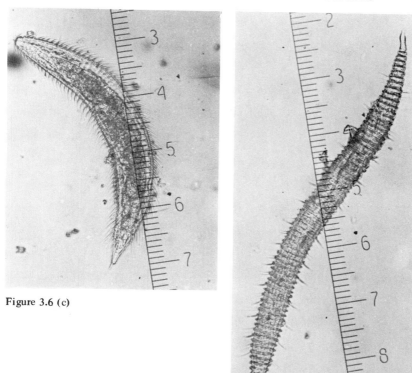

Figure 3.6 (c)

(d)

(c) and *(d)* Two genera of marine Desmoscoleicida free-living forms characterized by long cuticular projections of unknown function. These are marine species.

there are many which have localized paired pigment spots, usually embedded in the pharyngeal musculature. These have been called "pigment spots", "eyes", "eyespots" or "ocelli", although a photic response is reported for only one of these species, *Chromadorina bioculata*.

The ultrastructural organization of these photoreceptors has shown that in some, the conspicuous pigment is granular, probably melanin, and acts as a shading pigment for the small photoreceptive organelle (figure. 3.8). All the photoreceptors described so far are "rhabdomeric" receptors, the "rhabdomere" being a laminated or microtubular structure upon which the photolabile chemical is arranged. It has been reported that a nervous connection links the rhabdomere with the lateral cephalic nerve.

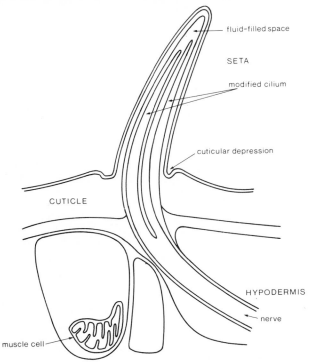

Figure 3.7. Reconstruction of sensory somatic seta, believed to be a mechanoreceptor, from the nematode *Chromadorina bioculata* (after Croll and Smith, 1974).

The photosensitive pigment in most species is probably a carotenoid. Some species in the genus *Enoplus* appear to have localized pigment spots, but these are not associated with a rhabdomere, and are probably not receptors but are accumulations of excretory by-products.

Mermis nigrescens, a parasite in grasshoppers in its larval stages, and living in the soil for many months after emergence from its host, has a bizarre adult female stage. This is non-feeding and is the only stage to have a diffuse red area at its anterior end, called the "chromatrope" and consisting of oxyhaemoglobin. If placed in the dark, *M.nigrescens* ceases to lay eggs, but resumes oviposition when illuminated. There is some evidence that the chromatrope is photosensitive and may control the rate of oviposition. If this is so, it is the only known case where haemoglobin is involved as a mediator in a photosensitive response; it remains unresolved. It has been proposed that the chromatrope may be sensitive to changes in

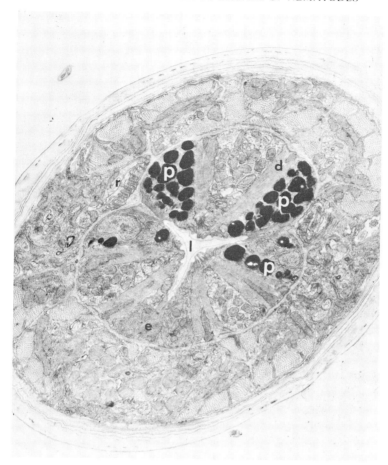

Figure 3.8. Electron micrograph showing a transverse section of *Chromadorina bioculata* through both ocelli. The orientation of the oesophagus (e) and lumen (l) shows that the two shading pigment cups (p) are dorso-lateral, between the dilator muscles (d), and they surround the rhabdomeric photoreceptors (r). (Croll, Riding and Smith, 1972).

oxygen concentration but, when placed in water saturated with carbon dioxide, nitrogen or carbon monoxide, there is no shift in the oxy-haemoglobin spectrum.

Many species of nematode show photosensitivity, and these are believed to receive light stimulation through a dermal light sensitivity, without localized receptors. Such familiar examples as *Amoeba, Hydra* and

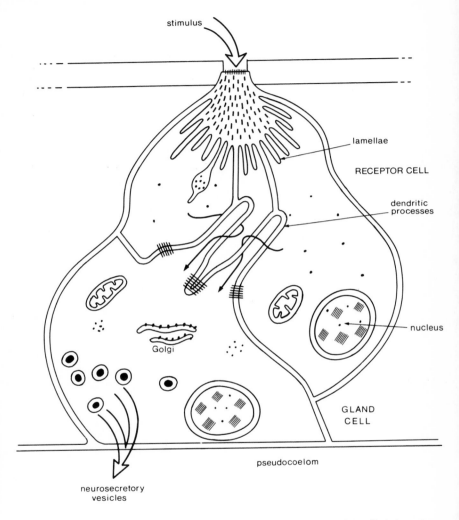

Figure 3.9. Diagrammatic reconstruction of the hypodermal gland cell of *Capillaria hepatica*. The importance of these cells is that they may be organized as a receptor and effector without nervous tissue. It is postulated that a stimulus enters in the pore on the surface. The effect is transmitted via "dendritic processes" to the effector gland cell. The effector gland cell releases neurosecretory vesicles into the nematode pseudocoelom (redrawn after Wright and Chan, 1973).

turbellarian flatworms perceive light through a dermal sensitivity. In molluscs and arthropods, direct photosensitivity of nerves has been reported. It may be that the transparent bodies of nematodes permit light to fall directly on nerves.

Spicules

The males of *Heterakis gallinarum*, *Nippostrongylus brasiliensis*, and *Heterodera* spp. all have spicules, through which runs nervous material. The nervous material contains cholinesterase and may be related to the spicule's function as a chemoreceptor or mechanoreceptor. The spicules, which hold open the vulva and transmit spermatozoa into the female during mating, may be sensory structures in all species.

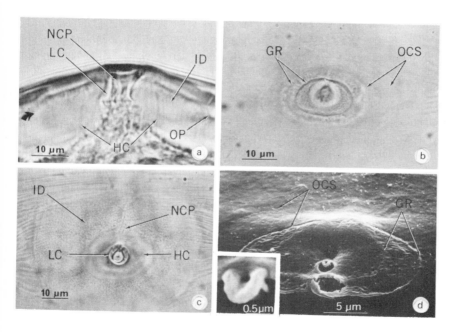

Figure 3.10. Ventro-median supplement of *Deontostoma timmerchioi. (a)* Left lateral view, anterior to right. *(b)* Ventral view anterior at top. *(c)* Ventral view focused at arrow in *(a). (d)* Scanning electron micrograph of ventral view with anterior at top. Inset is enlarged view of the organ protruding from the supplement (Hope, 1974). LC lining of canal. HC cylinder of homogenous cuticle. GR granular ring. ID disc-shaped region. OP outside perimeter of disc. OCS oval configuration of fine striae. NCP tube-like structure within the canal of the supplement and within the hypodermis.

Cuticular receptor-effectors

A gland cell and receptor cell have been found in association in the body wall of *Capillaria hepatica* (Wright and Chan, 1973). This is of particular interest as it appears to be one of the few receptor-effector cells of this kind yet found in nematodes. The sensory cell has a pit onto the surface of the nematode and is penetrated by "dendrites" from an effector cell adjacent to it. There is no nervous tissue found with these cells (figure 3.9). The adjacent cell contains vesicles identical to neurosecretory vesicles in nervous tissues. Wright and Chan (1973) postulate that changes in the receptor cell which occur in response to environmental input are monitored by the dendrites and lead to the release of neurosecretory materials into the pseudocoelom. The secretions may affect target cells at some distance from the effector cell.

Other sensory cells

Nematodes have a neural commissure, arranged as a girdle that goes halfway around the body on the ventral surface. It is called the hemizonid but its function is unclear (Smith, 1974). There are numerous small nerve

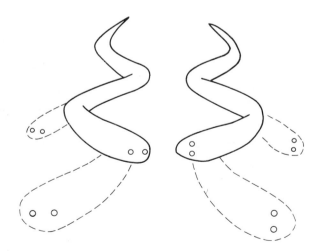

Figure 3.11. When nematodes swim, they propagate sinusoidal waves in two dimensions and swim on their dorso-ventral axis. Because the paired sense organs, such as the amphids, are situated laterally, they are actually dorso-ventral in movement. Therefore, a mechanism of environmental sampling must be derived from an arrangement such as that on the right and not that on the left.

cells throughout the bodies of nematodes which may be sensory cells. On the male tail of *Deontostoma timmerchioi,* a marine nematode from the Antarctic, is a structure called the ventro-median supplement (Hope, 1974). This new sense organ is belived to be a chemoreceptor. It is just anterior to the male cloaca on the ventral surface. The female of this species has large secretory cells along the lateral fields, and in copulation the male wraps its tail around the female. When the male is in copulation, the ventro-median supplement may obtain a stimulus from the female secretory cells which directs it to the vagina or initiates insemination (figure 3.10).

The reader will have seen that many of the references quoted in this section are from the early and mid 1970s; this field and the knowledge in it, are rapidly developing.

The problem of symmetry

In the earlier part of this chapter the arrangement of muscular hemicylinders was described as dorso-ventral and, because of this, nematodes usually move on their "sides" on a horizontal surface. Sense organs such as amphids, phasmids and ocelli have been described as "lateral", which means that during movement they are arranged one above the other (figure 3.11). In most bilaterally symmetrical systems it is thought that the bilateral arrangement permits alternate sideways sampling in movement. This apparent anomaly has not been solved in nematodes.

CHAPTER FOUR

THE BEHAVIOUR OF NEMATODES

THE TERM "BEHAVIOUR" IS VERY WIDELY USED AND MEANS DIFFERENT things to the different people who use it. Orientation behaviour was the most widely studied area in the first half of this century, but now attempts are being made to relate structure and physiology more closely to the activities and locomotion of nematodes. This account of nematode behaviour should be read in close association with Chapter 3 on nerves, muscles and sense organs of nematodes. The more modern approaches attempt to analyse the components of any action and to investigate nematodes as individuals rather than just as particles with speed and direction. The elegant and detailed cinematographic records of plant-parasitic nematodes during feeding are examples of such developments (Doncaster, 1971). Another major factor has been the application of tracking techniques, in which movements may be analysed accurately (Croll, 1975).

Below is a list of activities which have received specific attention and which are thought to involve specific sensory inputs, specific patterns of movement and to serve specific biological purposes (Table 4.1).

Many of these activities are linked in organized ways and use common movement components. In an attempt to blend the earlier knowledge and the more recent developments, some of the behavioural events will be discussed briefly with one or two examples.

The basis of activation

Under constant environmental conditions most nematodes undergo low-rate short-duration periods of movement. When feeding, they become less mobile and their pharynges pump in food. As pressure increases in the intestine, there is a spontaneous movement which causes defaecation. While feeding and moving, the uterus of the female undergoes spontaneous contractions which lead to oviposition. All of these activities may be grouped as "endogenous activities". They result from spontaneous neural

Table 4.1 Behavioural activities which are recognized in nematodes

Movement in the egg	Penetrating
Hatching	Nictating
Feeding	Leaping
Defaecating	Orientation to gradients
Moulting	Copulation
Exsheathing	Oviposition
Migration	Air breathing
Swarming	Shock reactions

or myogenic depolarizations, or are responses to turgor changes in the hydrostatic skeleton.

If an environmental change such as increase in light intensity, change in temperature, mechanical agitation, or a sudden change in the concentration of noxious or attractive chemical, is applied to nematodes, then a period of activity results. Different species will show various forms or durations of movement to the environmental changes, but the general pattern is widespread (figure 4.1).

The movement resulting from such environmental changes may be entirely non-directional, and to look only for directional orientations is to

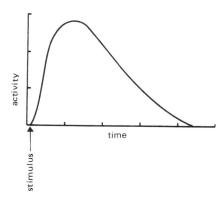

Figure 4.1. A generalized stimulus-response curve for nematodes. The stimulus is a change in the ambient conditions. The activity may be measured in terms of rate of movement, or the proportion of a population with movement. The duration of activity may be in seconds, minutes or hours, depending on the species and physiological conditions. This form of activity is classified as exogenous (Croll and Al Hadithi, 1972).

overlook the kinetic movements or changes in rates of activity which are common in nematode behaviour.

Gradual environmental changes or changes of insufficient intensity do not cause activation. Such activation is believed to result from external environmental input and is collectively recognized as "exogenous activity". Orientation responses to gradients must by definition result from exogenous input.

Activity may therefore result from exogenous or endogenous stimulation, or an integrated coordination of both. Nematodes do not do everything at once. Their behaviour is complex and patterned, and the mechanisms of control are being actively pursued. As a working hypothesis exogenous activities are believed to use exteroceptors, the central nervous system and the somatic musculature. Endogenous behaviour is coordinated by the sympathetic nervous system and the turgor changes.

Temporal and spatial patterns of movement

Small nematodes can inscribe their tracks in the surface of agar as they move over it. This provides a temporal and spatial record of movement. In over two dozen nematode species from widely diverse biological groups, repeatable features have been seen in tracks. Most waves are propagated backwards from the head and cause forward movements. These waves tend to occur at a fairly constant rate for any one species and any physiological status. The waves inscribe an asymmetrical path characterized by arcs and spirals which results from a series of waves which move further to one side than to the other (figure 4.2). When the organisms respond to gradients of attractants, the tracks tend to become straighter.

All nematodes stop spontaneously during movement and move backwards using waves propagated at the tail. These waves are often more rapid than backwardly directed waves, and are of short duration. "Reversals", as bouts of backward movement are termed, disrupt the pattern of tracks and lead to new directions being followed. They are also common in shock responses or in areas at the lowest and highest places in gradients of stimuli. The mechanism of neuromuscular integration causing arcs and spirals and the pattern of backward and forward waves in movement is not known. These are basic behavioural features in all nematodes from microfilariae consisting of a few cells and may be only 70 μm long to *Ascaris lumbricoides* which is up to 25 cm long.

Track analyses or other attempts to quantify behaviour have shown that nematodes are all individuals. One illustration of this phenomenon is given

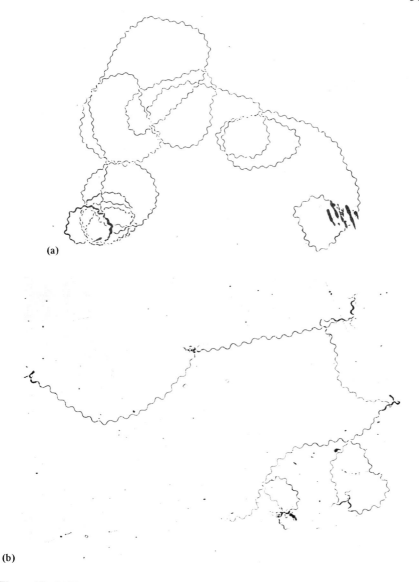

(a)

(b)

Figure 4.2. *(a)* Track of an infective larva of the cat hookworm *Ancylostoma tubaeforme* moving in a non-directional environment. *(b)* Track of an adult female *Caenorhabditis elegans* in a non-directional environment. Note the disruptive reversal bouts.

Table 4.2. Individuality of nematode behaviour. Values of three successive arcs (angle in degrees × radius in mm) in the tracks of ten individual hookworms (Croll and Blair, 1973).

Larva number	1st arc	2nd arc	3rd arc
1	35	42	48
2	32	12	24
3	30	25	14
4	25	48	27
5	20	12	6
6	16	16	12
7	14	6	8
8	12	8	8
9	8	8	8
10	4	7	4

in Table 4.2, in which ten infective larvae of *Ancylostoma tubaeforme* were tracked, and the parameters of their tracks were measured. As the larvae moved, they inscribed arcs or pieces of spirals, each with a radius and a circumference, measured as the angle subtended at its centre. The table shows the product of angle and radius for these thirty arcs. As can be seen, there is a highly significant individuality or (in statistical jargon) a significant autoserial correlation coefficient.

Movement in the egg and hatching

Nematodes begin to move as they complete their embryogenesis to the tadpole stage, but before elongation to a fully recognizable larval stage. The first movements are lateral uncoordinated "twitches" of the anterior end, but these soon develop into full body waves, and the larvae move actively within their eggs. Movement is known to be essential for the emergence of nematodes, even those that emerge within the gastrointestinal tracts of their hosts (figure 4.3).

Emergence from the egg is polar or subpolar. The mechanics of this has not been well documented, but presumably the larva is best able to exert pressure in this posture (figure 4.4). The most refined and coordinated

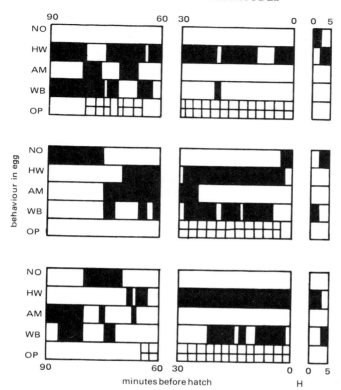

Figure 4.3. Movements of *Necator americanus* inside the egg before and immediately following hatching (H).
NO, no movement; HW, head waving; AM, anterior movement; WB, whole body waves; OP, oesophageal pumping (from Croll, 1974).

method of emergence from the egg is that of the potato cyst nematode *Heterodera rostochiensis* (Doncaster and Seymour, 1973). The stylet of this plant parasite is used systematically to perforate the egg shell by cutting a slit in it. Within 80 seconds the stylet has been seen to thrust 57 times. The extent of sensory feedback for this piece of behaviour is remarkable because, if the stylet fails to penetrate the eggshell on the first thrust, it persists at the same angle until it does. After penetrating the shell, it changes the angle of attack 1° at the fulcrum of its movement and this continues until a straight slit is drilled in the eggshell. The angular increment, straight line of cuts and delicate proprioceptive feedback is that of a most sophisticated system (figure 4.5).

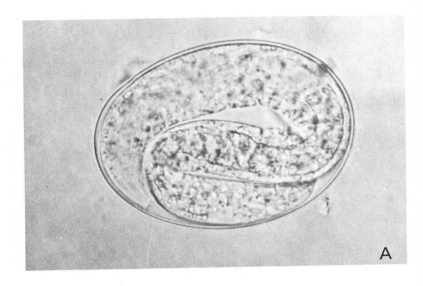

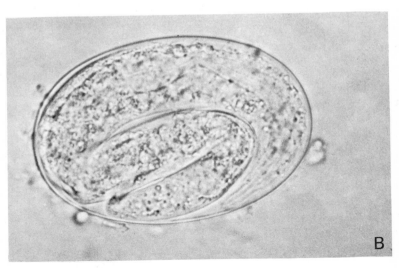

Figure 4.4. Two positions adopted by the human hookworm *Necator americanus* within the egg (Croll, 1974).

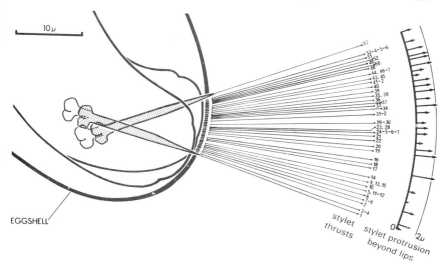

Figure 4.5. Diagram based on analysis of cine film of *Heterodera rostochiensis* larva cutting part of exit slit through the eggshell. Outlines of head and of stylet (protracted and retracted) at start and end of sequence shown. Within 80 seconds, 57 stylet thrusts were made; in 26 of these the stylet perforated the shell. Each arrow shows where a perforation occurred, and how far the stylet protruded beyond lips. As numbering of thrusts shows, when stylet failed to penetrate, it was thrust again (up to four times) nearer the previous perforation. After this sequence, the rest of the slit was made in the opposite direction, beginning near its original starting-point. Between these cuts an intact portion of shell remained which parted when the nematode emerged (Doncaster and Seymour, 1973). (E.J. Brill, Leiden)

Feeding

The ingestion of food has been filmed in a number of plant-parasitic and microbivorous species. The simplest feeding mechanism is in microbivores in which the triradiate lumen of the anterior pharynx (procorpus) is opened by the contraction of the pharyngeal radial muscles. Food is pulled back to the metacorpus, then to the pharyngeal bulb, and finally pushed under pressure back to the intestine (figure 4.6). This simplified form of ingestion is modified only in detail for ingesting blood, tissues, mucus and plant cytoplasm.

Control of feeding may be illustrated by the construction of a simple flow diagram. Figure 4.7 emphasizes that a patterned series of events occurs in a constant sequence, and that each depends for its initiation on the completion of the previous events. This flow is for a generalized plant feeding form such as *Aphelenchus avenae* which ingests the cytoplasm of fungal hyphae.

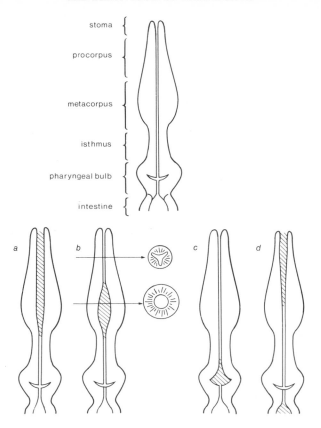

Figure 4.6. Pharyngeal structure of a typical microbivorous nematode. *(a)–(d)* sequence of ingestion of food.

A subroutine is included in the diagram, which permits for an increased rate of stylet thrusting when the cell presents a greater resistance. This type of analysis and multiphase nature of feeding has prompted Crofton to say:

> Students of ergonomics and system control might well consider the nematodes to be an original source for their studies.

In addition to the detailed muscular movements frequently emphasized in descriptions of feeding, it is characteristic that individuals become immobile while feeding. During ingestion, characteristic postures are assumed in which the body is consistently formed into a straight trunk with a curled or kinked tail. This indicates some form of communication, either

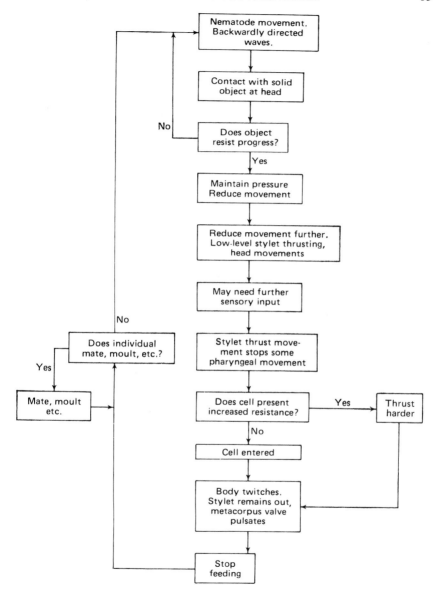

Figure 4.7. Flow diagram to show the sequential pattern of feeding characteristic of a simple fungal-feeding plant-parasitic nematode (Croll, 1975).

neural or hydrostatic, and must involve all the contractile elements of the body (see also Chapter 3).

Penetration

In addition to penetrating the egg shell and probing into host cells during feeding, "penetration" of tissue barriers is essential for the completion of nematode life cycles. Many animal-parasitic species undergo migrations around the body of their hosts before becoming established in the adult site. Such migrations may be extensive and involve passage through many barriers. Regretably, the behavioural aspects of these migrations are poorly known, but enzymes such as proteases, esterases and hyaluronidase have been implicated in penetration mechanisms.

The invasion and penetration of skin by hookworm larvae have received more attention than other aspects of nematode penetration behaviour. The surface of mammalian skin that is presented to invading larvae is composed of the dead keratinized cells of the stratum corneum, and is characterized by many folds and micro-lesions. It is probable that these provide entry sites for penetrating larvae. The skin surface is generally warmer than the environment and, in addition to a thermodynamic increase in activity, hookworm larvae show a positive thermotaxis (see page 74). These factors ensure that the larvae are most active when they contact the skin surface. Direct evidence for active enzyme production during penetration is lacking, but for some species indirect methods have demonstrated their presence. No enzymatic activity has been found for other species, and a mechanical penetration mechanism has been postulated. It appears that the details of penetration differ between these species. For those species which use pharyngeal secretions for penetration, but do not ingest the products of digestion, penetration may be considered as a modified form of feeding.

Air swallowing

A further way in which the stoma and pharynx may be involved in behaviour is in the novel proposal that it may pump gases into the intestine. The process has been observed in half a dozen or so separate species, in the microbivores *Pelodera*, *Rhabditis*, *Cephalobus*, *Panagrellus* and *Mesodiplogaster* and the marine *Metoncholaimus*. When *Mesodiplogaster lheritieri* were placed in situations where the gas phases could be changed, a significant increase in the swallowing of gas bubbles occurred in nitrogen, as compared with oxygen. If a gas mixture (80% N_2, 15% CO_2 and 5% O_2)

was used instead of nitrogen, a greater increase in air swallowing resulted. Klingler and Kunz (1974) discussing these results felt that the low O_2 and higher CO_2 stimulated the greatest rate of air swallowing. The bubbles pass back into the intestine, where they often coalesce and soon disappear. These records have the added interest of providing the phylogenetically earliest "breathing" mechanism of any animal group.

Defaecation

During feeding, the movements of the hind gut cause the egestion of fluid—the process of "defaecation". The squeezing of fluid from the rectum and anus is a carefully controlled process because of the internal hydrostatic pressure. In *Ascaris lumbricoides*, fluid is ejected with such force that it can reach three feet across the laboratory—longer than arms length! Defaeca-

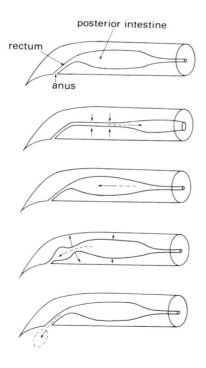

Figure 4.8. Stages in the defaecation of *Caenorhabditis elegans*. This series illustrates an example of localized muscular contraction, causing displacement of fluid.

tion is closely integrated with hydrostatic turgor changes in the nematodes; after ejection of fluid in *A. lumbricoides* there is a contraction of the whole body to compensate for the loss of fluid.

It is possible to observe the movements of fluid inside *Caenorhabditis elegans* during feeding and defaecation, and fluid is ejected about every 100 seconds. Firstly, the posterior intestine gradually dilates as the pharynx pumps food into it (figure 4.8). There is then a sudden localized contraction in the hind intestine which forces fluid forwards. A contraction of the whole body forces the intestinal contents back into the hind intestine and dilates the rectum. The rectum is then finally evacuated.

Nictating and swarming

The infective larvae of some species are able to mount projections and, while supporting themselves by their posterior tip, can wave in three-dimensional spirals and loops, a process known as "nictation". Some species can exploit the meniscus forces of such situations to coil their bodies into a bow and "leap" for distances many times their own length. In drying microhabitats, hookworm infective larvae may break through the anterior tip of their sheaths and extend their bodies out of the sheath, waving from side to side. If the substratum is tapped or there is some similar sudden change, the larvae retreat rapidly into their sheaths. Such movements must require considerable muscular coordination.

Figure 4.9 (a)

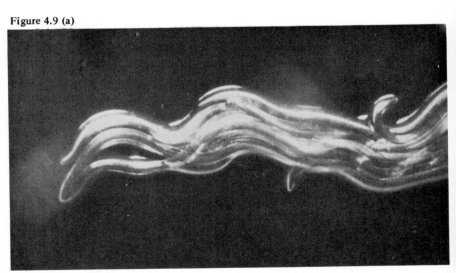

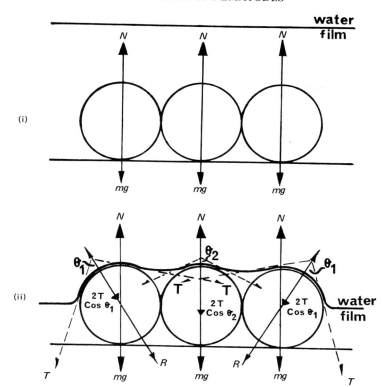

Figure 4.9. *(a)* Swarm of *Panagrellus redivivus* *(b)* Summary of the external forces acting on *(i)* three individuals of *Panagrellus redivivus* at rest in a body of water and *(ii)* in a film of water.

T, surface tension; m, mass; g, acceleration due to gravity; N, forces acting at right angles to the substrate. $2T \cos \theta = R$ are the forces acting on the outer worms that by their inward direction tend to keep the swarm together, so that the worms act as a unit (Croll, 1970).

In dense populations of nematodes moving in thin fluid films, groups of individuals become synchronized and form "swarms". Morphological alterations in the cuticle have been claimed for some plant-parasitic forms, but in *Panagrellus redivivus* the meniscus provides the mechanical forces, and no specific behavioural actions need be invoked (figure. 4.9). This form of swarming is species and stage-specific, because all the members of the swarm must share common movement characteristics. Larvae of the same species cannot maintain their position in a swarm of adults, and they establish larval swarms.

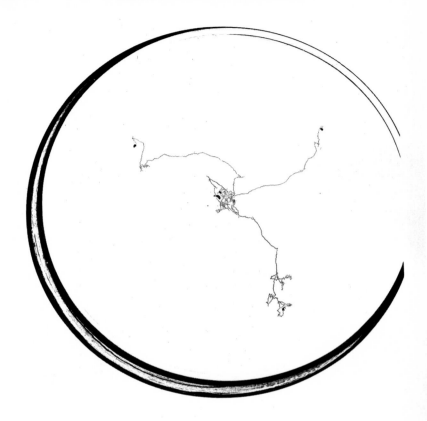

Figure 4.10. Tracks in agar of three individuals of *Caenorhabditis elegans* responding to gradients of NH₄Cl applied at the centre (Ward, 1973).

Orientation responses

There is a large and widely dispersed knowledge of the subject of nematode orientations, and most of the earlier studies on "nematode behaviour" were investigations into orientation responses. Evidence for the sensitivity of nematodes to selected stimuli will be briefly listed with one or two examples of each. For a fuller account of work up to 1969 see Croll (1970).

Chemosensitivity and responses

Nematodes can respond to chemicals in many different situations and the amphids, phasmids, spicules and papillae are probably all chemoreceptors (p.43 et seq.). The anions Cl^-, Br^- and I^- and the cations Na^+, Li^+, K^+, Mg^{++} were all found to be attractive to *Caenorhabditis elegans.* Alkaline pH causes the accumulation of *C. elegans,* as do cyclic AMP and cyclic GMP (figure 4.10) (Ward, 1973). Sugars and amino acids have been shown to be attractive to many species in suitable concentration gradients. At high concentrations of "attractants" there is usually a repellent effect, e.g. *Ditylenchus dipsaci* was attracted by 1:100,000 glutamic acid and aspartic acid, but repelled by 1:1000.

Carbon dioxide has been implicated in many behavioural activities, either as a gas or as undissociated carbonic acid; its role has developed into something of a universal attractant. Carbon dioxide stimulates larvae of

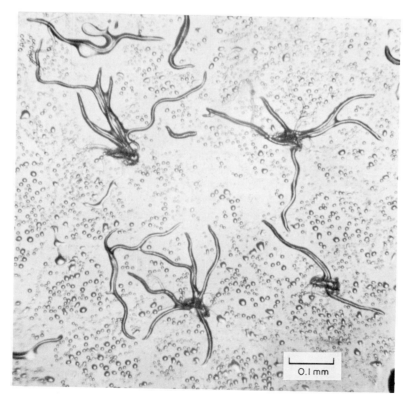

Figure 4.11. *Aphelenchoides fragariae* penetrating through holes, out of which a thin stream of carbon dioxide is passing (Klingler, 1970).

Ascaris lumbricoides to emerge from the egg when presented in the correct physico-chemical micro-environment. Exsheathment of infective larvae of *Haemonchus contortus, Nematospiroides dubius* and other trichostrongyles is also stimulated by carbon dioxide. Hookworm larvae are stimulated to mount projections and nictate, waving from side to side at the carbon dioxide concentrations of human exhaled breath.

The plant-parasitic nematodes *Ditylenchus dipsaci* and *Aphelenchoides fragariae* attack the stems and aerial parts of plants, and move towards microjets of carbon dioxide. The biological role of carbon dioxide in the life of *A. fragariae* was tested by Klingler (1970) who made perforated plastic

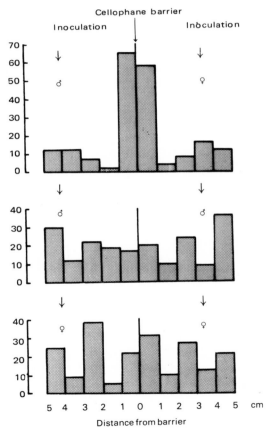

Figure 4.12. The movements of *Panagrolaimus rigidus* in tubes with homosexual and heterosexual combinations on either side of a cellophane barrier (redrawn from Greet, 1964).

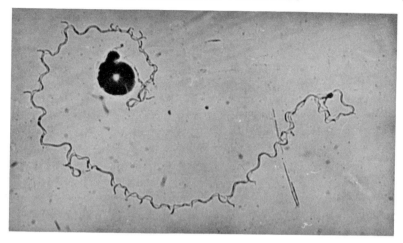

Figure 4.13. The track of a male *Heterodera rostochiensis* as it approaches and then finds the sedentary female (Green, 1966).

membranes over which the nematodes moved freely. When carbon dioxide was passed through the pores in the membrane at rates comparable to those occurring through the stomata and lenticels of living plants, the worms rapidly accumulated at the pores (figure 4.11) and entered them.

Gradients of carbon dioxide occur around the roots of plants, in the rhizosphere, up which root parasites travel when locating their hosts. Carbon dioxide is also generated in such biological situations as decomposing organic debris and by metabolically active plants or animals. While it is a relatively non-specific cue, it is widely used as an environmental signal to locate hosts and areas of food. In a similar context the ammonium ion is attractive to the microbivorous nematode *Rhabditis oxycerca*. Ammonium ions are commonly produced during decomposition and may also occur around roots because of the ammonifying bacteria.

Pheromones are substances which, when produced by an organism and secreted into the environment, influence the behaviour of other members of the same species. Such substances occur in nematodes and are known largely from studies on sex attractants (figure 4.12; 4.13). When females and males of *Panagrolaimus rigidus* are separated by a cellophane barrier, both sexes move towards members of the opposite sex (Greet, 1964). No movement occurs when males and males, or females and females, are used (figure 4.12). This was the first demonstration of sex attraction in nematodes, and the phenomenon has now been confirmed in many diverse nematode groups. The most detailed account of the behaviour involved in sex attraction is that of Green (1966). He examined sex attraction of males to the sedentary females of the cyst nematodes *Heterodera* spp. In this very complex picture, at least six different male attractants occurred in the genus *Heterodera* (figure 4.13). Of the ten species tested, most females secreted more than one attractant, and most males responded to more than one attractant. The attractants have not been chemically defined, but they are water-soluble, non-volatile, and diffuse through dialysis membranes.

In *Pelodera strongyloides* the male is attracted to both sexes, while the female is attracted only to the male, and the attractant may be blocked with phosphate buffer. On this and supporting evidence, Stringfellow (1974) claimed that the attractant may be the hydroxyl ion. Sex attractants are effective only with adults or late fourth-stage larvae.

Orthokinetic activation, microfilarial periodicity, feeding stimuli, hatching and exsheathing stimuli, host location and copulation are just some of

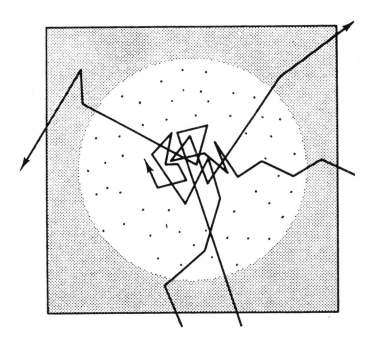

Figure 4.14. The photoklinokinesis of infective larvae of the horse strongyle *Trichonema* sp. Note the increased rate of turning as the larva enters illuminated area. Heavy stippling 5 ft-c, light stippling 300 ft-c, plain 900 ft-c. (Croll, 1965).

the behavioural activities in the lives of nematodes in which chemical stimulation is implicated.

Photosensitivity and responses

Many experiments have shown the accumulation of nematodes at one or other end of a light intensity gradient, but too often the heat or infra-red portion of the light has not been adequately removed. Those nematodes which respond to directional "cold" light are photopositive at intensities of 100–1000 c/ft². These photo-responses are not necessarily phototaxes, as has often been assumed, but are probably photoorthokineses or photoklinokineses. Thus worms are activated by increased incident illumination, or

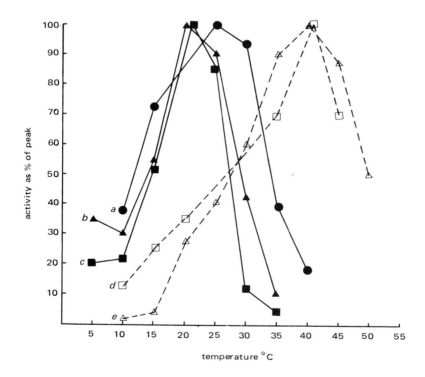

Figure 4.15. The rate of activity of infective larvae of nematodes at different temperatures. Continuous lines, plant parasites and larvae of ruminant parasites which enter on herbage during grazing; dotted lines, active penetrators of warm-blooded hosts. *(a) Ditylenchus dipsaci; (b) Tylenchorhynchus icarus; (c) Trichostrongylus colubriformis; (d) Strongyloides ratti; (e) Ancylostoma tubaeforme*

their rate of turning or reversing is changed by changes in light intensity (figure 4.14). The result of such kineses can lead to the accumulation of nematodes at the bright or dark end of a gradient, but not through a phototaxis. As most species do not have localized photoreceptors, but instead are sensitive through a dermal light sense, kinetic responses are more likely than taxes.

The adult female of *Mermis nigrescens* which emerges from soil to lay eggs has a "chromatrope", which is believed to be a photoreceptor (see page 48). Many free-living marine and freshwater species have paired areas of pigmentation localized in the pharyngeal musculature (figure 3.8). These structures are photoreceptors but to date only one species *Chromadorina bioculata* has been shown to be photosensitive, and it is positively phototactic.

Thermosensitivity and responses

Temperature has a directly thermodynamic effect on metabolic rates and on rates of movement. There is a small temperature range over which each species moves most quickly. It is biologically noteworthy that actively penetrating infective larvae of warm-blooded hosts, such as *Strongyloides ratti* and *Ancylostoma tubaeforme* are most active around 35–40° C, while infective larvae of plant parasites or pasture forms that enter contaminatively on herbage are generally most active around 20–25° C (figure 4.15).

Adult nematodes from both warm-blooded and cold-blooded hosts move towards the hot end of a thermal gradient: these include *Nippostrongylus brasiliensis,* (rat), *Rhabdias bufonis* (frog) and *Camallanus* sp. (turtle). A striking feature is that they continue to go towards the hot end until they die from thermal damage. These have been described as thermotaxes, but only in hookworm larvae has the mechanism of the heat response been sufficiently examined to use the term "genuine thermotaxis".

When infective larvae of the cat hookworm *Ancylostoma tubaeforme* are placed in a "two-heat" experimental design, in which the larvae are released at right angles to the pair of stimuli, they inscribe T or Y-shaped paths (figure 4.16). A heat contour is established between the heat sources, such that 1 cm between them there is a temperature of 35° C, while 1 cm in any other direction the temperature is 30° C. Larvae following a thermotaxis cross the steepest gradients from a heat source, and so the pattern is as in figure 4.16 (Croll and Smith, 1972).

It has been shown that, amongst the plant parasites, some will aggregate at an eccritic temperature or thermal preferendum. This is not at the hot or cold end, when accumulation could be orthokinetic through hot or cold inactivation. The thermal preferendum may be changed by storage at a constant temperature for a month. Thus after acclimatization at 10°, 20° and 30° C, *Ditylenchus dipsaci* will show a significant tendency to accumulate at these temperatures (figure 4.17). The mechanism of acclimitization is unknown, but similar phenomena in other animals are based on mitochondrial or neuromuscular changes.

Mechanosensitivity and responses

If a living individual of *Ascaris lumbricoides* is held over a bench and dropped from about a foot, it will respond immediately by contracting all its somatic muscles. Strips of somatic muscle will show spontaneous rhythmic contractions when they are loaded with weights of 10–20g. Nematodes can be activated by touching them, and an imposed mechanical distortion of their bodies causes a change in the wave form after release. The muscle cells of nematodes respond directly to mechanical input and

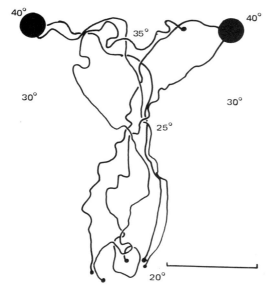

Figure 4.16. Tracks of infective *Ancylostoma tubaeforme* larvae when placed in a "two-source" heat gradient. The dark spots are 40° C heat sources; the scale is 1 cm. The way that the larvae first move between the sources and then to one or the other is typical of a thermotaxis (Croll and Smith, 1972).

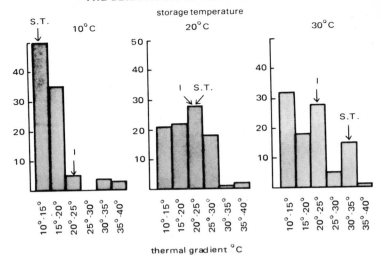

Figure 4.17. The distribution of *Ditylenchus dipsaci* in a thermal gradient following storage at 10°C. 20°C and 30°C. I = inoculation point, S.T. = Storage temperature. (redrawn from Croll, 1967).

hydrostatic changes. During hatching, feeding, penetration, mating and oviposition, an awareness through tactile sensitivity is implicit. There are many sense organs including setae, papillae, hemizonids, and proprioceptors which provide localized sensory input.

Only one quantitative investigation has been attempted to analyse mechanosensitivity (Croll and Smith, 1970). Tiny pins were dropped from measured heights using capillary tubes and electromagnets to release the pins. The tubes directed the pins onto specific regions of the bodies of *Rhabditis* sp. The exact kinetic impact could be measured, and this could be related to the form and duration of the response (figure 4.18). If stimulated at the anterior end, the worm moved backwards; if stimulated on the posterior half of the body, the worm moved forwards. The whole body musculature was able to respond in a completely integrated way to localized stimulation. This strongly suggests the involvement of the central nervous system.

It is likely that, together with chemosensitivity, mechanosensitivity is a fundamental property of all nematodes.

Galvanosensitivity and responses

Plant root surfaces develop negative charges, and these potentials have

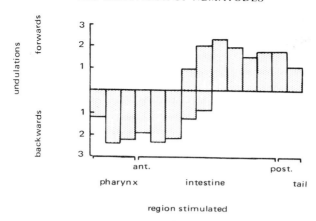

Figure 4.18. The response of *Rhabditis* sp. to mechanical stimulation of an equal intensity in different parts of the body. When stimulated in the anterior half, the worms move backwards; when stimulated posteriorly the movement is forwards (Croll and Smith, 1970).

been found to be about 60–100 mV in some cases. If placed in electrical potentials of the same order, about a dozen species of nematode have been shown to migrate to one or other pole. In fact the threshold for the response may be as low as 30mV. Most of the plant-parasitic species moved to the cathode, except for second-stage larvae of the sugar beet nematode *Heterodera schachtii*. Larvae of *Trichostrongylus retortaeformis* moved to the cathode of an agar bridge in currents of 10–40 mA. Microfilariae of *Litomosoides carinii* in serum-tyrode and larvae of *Pelodera strongyloides* migrated to the anode, whereas adults of *P. strongyloides* were unaffected (Croll, 1970).

The sensory basis of galvanotaxis, if any, is obscure, but protozoans, molluscs, lampreys, goldfish and others all move directionally in electric potentials. The immediacy of the response to a change in the direction of the current, and the size of the organisms responding, eliminates such possibilities as response to products of electrolysis or passive electrophoresis.

Geosensitivity and responses

Gravity does not effectively change in its direction or intensity, and so all responses must be taxes as, by definition, "kineses" require changes in the intensity of stimulation. The upward migrations of infective larvae on pasture herbage, or of hookworms in soil, led to the early assumption that such forms are responsive to gravity. When spreading randomly from a point source, some larvae will indeed move upwards, but whether this is significant is not clear. Surface films acting on such larvae exert a force 10^4–10^5 times greater than gravity, which helps to place gravity in a wider context. There are many claims and counter-claims for geotaxes, but a consensus view is not yet available.

Turbatrix aceti is tail-heavy, and this drag explains the consistent upward movement of *T. aceti* in bodies of fluid. The absence of suitable sense organs may also suggest a lack of sensitivity, but the same would apply to miracidia, cercariae, and others with reported geosensitivity.

General statement on nematode behaviour

Behaviour varies between species and between different stages in the life of a single species. This is most dramatically illustrated among the parasites which have bacterial-feeding preinfective larvae, non-feeding infective larvae and parasitic adults. Each phase has its own specific behavioural responses. In addition to behavioural changes, sense organs are known to alter during the life cycle; this is related to moulting. The hydrostatic pressure attributed to nematodes provides not only a force antagonistic to the longitudinal muscle in movement, but also a means of endogenous proprioceptive communication. This probably plays a part in movement, feeding, defaecation, copulation, oviposition, and hatching. There is a division of behaviour into endogenous or exogenous activities, or an integrated mixture of both. Endogenous activities are spontaneous, autonomous, and include posture maintenance, oviposition and feeding. Exogenous actions result from external input, are largely cholinergic, are rapid, and have hierarchical dominance over endogenous components.

Neuromuscular and sensory organization is a mixture of radial and bilateral symmetry, and a first-order mechanism in which the receptors achieve orientation through alternate lateral sampling is too simplistic.

Nematodes respond to separate input modalities; they have endogenous pacemakers, and a complex range of behaviours which at times show considerable refinement and complexity. *Caenorhabditis elegans* has less than 300 neurones and, if this is a representative figure, it is within this neural capacity that we must contain nematode behaviour.

CHAPTER FIVE

THE FEEDING AND NUTRITION OF NEMATODES

MOST NEMATODES FEED THROUGHOUT THEIR LIFE CYCLES TO PROVIDE energy for basal metabolism, growth and reproduction. Certain stages, however, are non-feeding. The most common examples occur among the free-living infective stages of parasites, e.g. the second-stage larva of *Meloidogyne hapla,* or the third-stage larvae of *Ancylostoma duodenale* and *Nippostrongylus brasiliensis.* Special physiologically distinct stages called "dauer" larvae (from the German word meaning "waiting") can develop in free-living rhabditids in response to poor environmental conditions. These long-lived resistant non-feeding stages share many characteristics with infective stages. *Mermis nigrescens* is parasitic in insects in its larval stages where it feeds and grows, but adult *M. nigrescens* live in the soil for over a year and do not feed.

Many animal-parasitic nematodes, microbivorous soil forms, and marine forms live in their food. Their nutritional demands must therefore satisfy both the chemical feeding requirements and the physical requirements such as texture, pH, redox potential and temperature.

The basic nematode is a tube within a tube. The inner tube is the gut. It consists of a stoma, a pharynx, intestine, rectum and anus (figure 5.1). The stoma may be simple and undifferentiated, as in *Ascaris lumbricoides,* or it

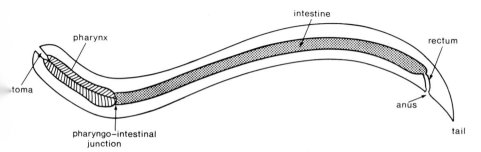

Figure 5.1. Generalized nematode gut.

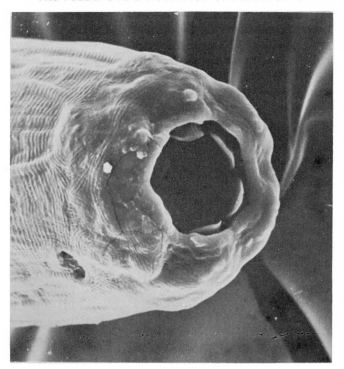

Figure 5.2. Stereoscan micrograph of the stoma of the predatory nematode *Prionchulus.*

may be a simple cylinder as in the bacterial feeders, *Rhabditis, Caenorhabditis* and *Pelodera.* In plant-parasitic forms the stoma has been modified into a stylet or spear. This is used to pierce the host cells and pump back the cell contents. Hookworms have large plates and "teeth" to hold and tear host mucosal tissues. Predatory nematodes like *Iotonchus* and *Mononchus,* which feed on nematodes and other metazoans, develop a shearing force against their prey, using stomatal plates and a powerful suction in their pharynx (figure. 5.2). Generally speaking, an experienced nematologist, upon being presented with an unknown nematode, could say a great deal about its feeding habits after an examination of its stoma. In the same way, a mammalogist could describe the diet of a mammal after inspecting its dentition.

Behind the stoma is the pharynx or oesophagus, which may be modified, but is basically a muscular pump with valves and a complex contractile organization. Associated with the pharynx are three pharyngeal glands, the ducts of which empty into the pharyngeal lumen. The function of the pharynx is to pump food into the intestine against the hydrostatic pressure, to lubricate it, and to mix enzymes with it.

During ingestion, the anterior pharynx of most tylenchid nematodes is dilated by the contraction of muscles, and fluid is sucked in (figure 5.3). This is followed by the dilation of the lumen of the pump and posterior pharynx, and closure of the anterior pharynx. If the pump contracts, an inward movement of fluid results, and the fluid enters the intestine. Doncaster (1971) has considered the theoretical mechanics of this system. As the pharynx is pumping food against a pressure gradient, it is not immediately obvious how a pump with only one outlet valve is sufficient to explain unidirectional flow. To rephrase the problem: When the pump contracts, why should the food pass backwards and not forwards? (figure 5.3).

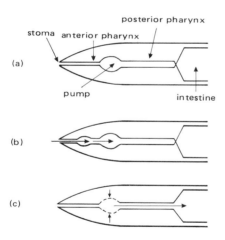

Figure 5.3. Mechanical changes in the hypothetical pharynx (see text) *(a)* Generalized components. Note that the radius of the anterior pharynx is much smaller than that of the posterior pharynx. *(b)* The pharyngo-intestinal junction remains closed as the anterior pharynx dilates to prime the pump. *(c)* When the pump contracts, fluids pass backwards because of the greater radius of the posterior pharynx.

The positions of the valves and the diameters of the tubes become all-important in these designs. Poiseuille's Law states that the volume of liquid Q, which flows per second along a capillary tube, through a pump under a pressure P, is related to the length L and the radius R of the capillary, and to the viscosity η by the equation

$$Q = \frac{PR^4}{8L\eta}$$

The over-riding position of the radial measurement is clear from this equation. If in a hypothetical nematode the pump is midway along the pharynx, having a radius of 0.08 μm in front of the pump and 0.23 μm behind it, then the resistance to "back flushing" along the anterior tube to liquid flow will be $0.23^4/0.08^4 = 68.32$ times that of the posterior tube. Fluid will therefore pass backwards because of the radial diameter.

Food passes backwards into the intestine for digestion and absorption, and the egesta is passed to the rectum and out of the anus. The rate at which food passes along the total alimentary tract varies widely, but during active feeding the fastest are probably *Ascaris lumbricoides* which evacuates its intestine about once every minute and *Caenorhabditis elegans* once every ten minutes.

Digestion and absorption of nutrients

The intestine of nematodes is the main, and in most species the only, absorptive surface for nutrients. Food is digested extracellularly in the lumen by secretions from the intestinal cells or by the pharyngeal gland secretions which mix with the food as it passes backwards. In some plant-parasitic nematodes there is extracorporeal digestion, discussed below.

The oxyurid nematode *Cosmocerca ornata* feeds on bacteria, partially digested food and cellular debris in the hind gut of its host, the frog *Rana temporaria*. Colam (1971a) consistently found *C. ornata* in the host's food, and concluded that the nematodes migrated along the intestine in the food. Digestion of food in the intestine of *C. ornata* was entirely extracellular. Another "pinworm", *Aspiculuris tetraptera,* lives on bacteria and cellular debris in the hind gut of mice, and digestion is also extracellular. These pinworms, and the human pinworm *Enterobius vermicularis,* are non-pathogenic and cause disease only if the populations of parasites are dense enough to cause blockages. Their lack of pathogenicity is related to their biologically-harmless feeding habits.

The nematode intestine is one cell thick and has been examined in a large number of species. The inner luminal surface is covered in absorptive microvilli, and the outer pseudocoelomic surface is bounded by a basement membrane (figure 5.4). The intracellular inclusions include mitochondria, lysosomes, dense endoplasmic reticulum, Golgi complexes, secretory vesicles and laminated degradation products. All of these are suggestive of

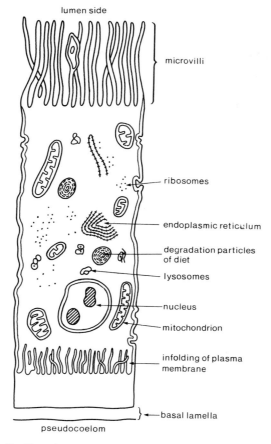

lumen side

microvilli

ribosomes

endoplasmic reticulum

degradation particles of diet

lysosomes

nucleus

mitochondrion

infolding of plasma membrane

basal lamella

pseudocoelom

Figure 5.4. Generalized intestinal cell of a nematode. The cell is clearly adapted for digestive, secretory and absorptive functions.

active cells involved in synthesizing and secreting proteins, and absorbing and partially digesting food. The microvilli of some species are branched and irregular in shape, and probably pass enzymatic secretions into the intestinal lumen. This type of secretion is known as "micro-apocrine" (Jenkins and Erasmus, 1969).

In the infective larvae of *Nippostrongylus brasiliensis,* Lee (1968) found intense staining for esterases in the anterior third of the intestine. There were also small widely-spaced microvilli. This reduction of microvilli correlates interestingly with the absence of feeding in the infective stage.

Lipid is the main food reserve of these infective larvae, and the demonstration of intense esterase activity is probably related to the mobilization of the lipid reserve. The anterior part of the intestine is thought to be mainly secretory and digestive, while the microvilli of the posterior part are longer and more dense, indicating an absorptive function.

When *Ascaris lumbricoides* was tied at the mouth and anus, and placed in labelled glucose or inorganic phosphate, no significant increase in these nutrients could be detected in the worms; whereas those individuals not tied, took up glucose and phosphate very rapidly. In discussing this result and the results of similar experiments, Read (1966) concluded that the cuticle could not absorb significant quantities of nutrient, and that the intestine was the main absorptive surface.

Active transport of glucose by the intestine of *Ascaris lumbricoides* has been demonstrated from one extracellular fluid phase to another, i.e. from the lumen to the pseudocoelom. The pseudocoelomic fluid of *A. lumbricoides* is rich in the disaccharide trehalose, and it has been postulated that the monosaccharide glucose is quickly converted to trehalose in the pseudocoelom. In this way a chemical gradient of glucose would be maintained, facilitating the transport of further glucose. Glucose can, however, be absorbed against a concentration gradient. The anthelminthic dithiazanine specifically blocks the uptake of glucose in *A. lumbricoides* and other nematodes, further suggesting specific sites for glucose absorption.

Amino acids in one study entered the intestinal epithelium of *A. lumbricoides,* but they remained in the cells. In another investigation Read (1966) used C^{14}-labelled amino acids, and reported that the intestine of *A. lumbricoides* took up histidine, glycine, methionine and valine at nonlinear rates. While being cautious about interpreting this result, it seems that some form of modified transport process exists for amino acid uptake, and that the mechanism shows some chemical specificity.

Harpur and Popkin (1973) developed a technique for using isolated intestines of *A. lumbricoides* for uptake experiments. About 15 cm of the posterior portion of the intestine were removed and tied onto a catheter; it was then rinsed thoroughly. It could be tied at the other end for a non-everted sac or, following the classical "everted sac" technique of vertebrate physiology, it could be turned inside out. A thin length of polythene was passed through the catheter and through the intestinal lumen. It was then tied onto the intestine, which was drawn back inside itself and through the catheter. After being tied at either end, the intestine could be placed into test solutions.

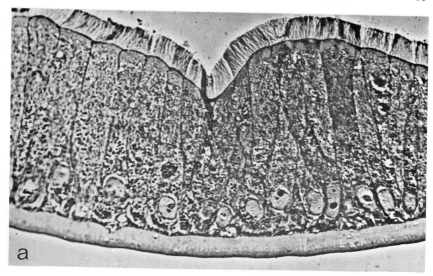

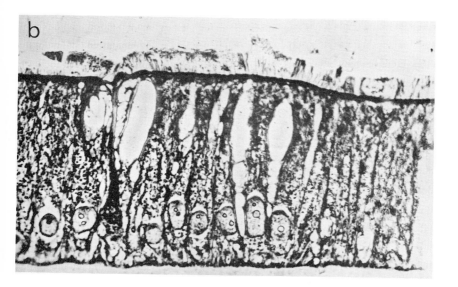

Figure 5.5. Sections of everted sac of *Ascaris* intestine *(a)* not transporting water, *(b)* after transporting water for ½ h (Harpur and Popkin, 1973).

When bathed in water, the everted sacs expanded significantly due to the net movement of water from the lumenal to the pseudocoelomic side. Water moved freely in osmotic gradients in both directions across the intestinal epithelium. A striking result of this study was the appearance of large intercellular spaces which developed between the columnar epithelium, and were presumed to dilate during the transport of water (figure 5.5).

In addition to these studies into the transport of water, glucose, and amino acids, the absorption of vitamin B_{12} by the gut of *A. lumbricoides* has been considered. Vitamin B_{12} (cyanocobalamine, the anti-pernicious-anaemia factor) is known to be absorbed in considerable quantities by such helminths as the tapeworm *Diphyllobothrium latum*. This and other helminths can cause vitamin-deficiency symptoms in their hosts. *Ascaris lumbricoides, Ancylostoma caninum, Ascaridia galli, Strongylus vulgaris, Toxocara canis* and *Trichuris vulpis,* all contain vitamin B_{12} and it may be that many more nematodes contain it. When incubated in high-B_{12} media, the coelomocytes of *Nippostrongylus brasiliensis* become reddish. *A. lumbricoides suum* from pigs has 3 to 4 times more B_{12} than *A. lumbricoides* from man; this is probably related to the differences in the bacterial flora of the host intestines. *A. lumbricoides* absorbs B_{12} most in the anterior part of the intestine (Zam and Martin, 1969).

Despite the relatively large amount of work on absorption by the intestine of *A. lumbricoides*, there is little knowledge of its foodstuff. Adults of *A. lumbricoides* cannot be considered to be site-specific, as they move throughout the intestine and bowel. When fixed and sectioned, the buccal cavity and intestine of *A. lumbricoides* contained a higher concentration of host intestinal cells than did the pig intestine (Davey, 1964). This indicated that it was either ingesting them preferentially or concentrating them in its intestine. There is no doubt that they ingest large quantities of undigested and digested food, bacteria and other debris.

As is often the case in nematode physiology, work on *A. lumbricoides* predominates, and the applicability of these restricted results to the phylum in general must always be questioned. The cuticle of *Ascaridia galli,* for example, is reported to absorb both glucose and alanine. Nevertheless, on the grounds of comparative morphology, it seems certain that the intestine will be shown to be the main absorptive surface in nematodes.

A variety of nematodes live in the haemocoel of insects; they show bizarre features of structure and biology. *Bradynema* sp. lives in the dipterous insect pest *Megaselia halterata;* it grows and reproduces in the haemocoel, but loses its feeding apparatus and gut. The cuticle of *Bradynema* sp. has been examined with the electron microscope, and was found

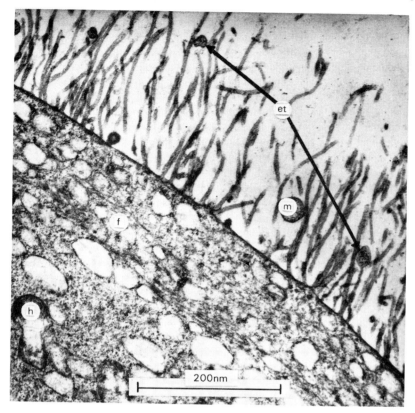

Figure 5.6. Transverse section of the body wall surface of *Bradynema* sp. showing microvilli (m) on the cuticle, some with enlarged tips (et) and supported by fibres (f) below in the hypodermis (h) (Photograph courtesy I.L. Riding).

to be covered in microvilli (Riding, 1970) (figure 5.6). This was the first study to provide evidence that the cuticle of this category was specialized to be an absorptive surface. Subsequent studies have confirmed the absorptive properties of the cuticle in biologically related nematodes.

The intestine lacks an organized muscular system, although there may be small intracellular contractile units. The movements of the intestine are generally considered to be indirect; food is pumped in at one end and flushed out at the other. Bodily movements also cause mixing and movements of the intestinal contents. Crofton (1971) was most dogmatic on this point:

The intestine is, of course, covered with pseudocoelomic membranes, and sometimes there are connections between the body wall musculature......These transpseudocoelomic fibres can give the spurious appearance of peristalsis but the only movement of the contents of the intestine, apart from filling and emptying, is by fluid transfer under the action of the waves of contraction in the body musculature.

One report by Lee and Anya (1968), who carefully observed living individuals of *Aspiculuris tetraptera*, indicated that the intestine could undergo coordinated contractions. They reported frequent peristalsis-like movements along the whole intestine, which were initiated anteriorly and were dissipated posteriorly. This report stands out as being somewhat unusual. Nevertheless it should be emphasized that there have been very few direct observations on nematodes other than plant parasites, which are all small and are all non-particulate feeders. Most conclusions about the intestine, including those of Crofton, are based on morphology alone. More direct observations of feeding in a wider spectrum of nematodes are needed.

Essential foods

Caenorhabditis briggsae feeds on bacteria in nature, but can be easily cultured on nutrient media of soy-peptone, yeast extract and fresh liver extract. By the systematic withdrawal of components and the specific supplementation of others, a picture has emerged regarding the essential nutrients which the nematode is not able to synthesize. The nutritional requirements are similar to those of other invertebrate metazoans.

Essential amino acids for *C. briggsae* are arginine, histidine, isoleucine, leucine, lysine, methionine, phenylalanine, threonine, tryptophan and valine. These are not synthesized at levels sufficient to permit reproduction. Alanine, asparagine, cysteine, glutamine, glycine, proline, serine and tyrosine can be synthesized.

It has been found that a dietary sterol, a haem, and a protein "growth factor", are dietary essentials; Vanfleteren (1974) added various organic and inorganic "carriers" for haem and found that *C. briggsae* could maintain its growth and reproduction. The haem component must be provided in a particulate form, which suggests that it is taken up in the intestine by phagocytosis. The protein "growth factor" may not be nutritionally essential, but it may serve only as a carrier for haem. If this is correct, the essential amino acids, a sterol and a haem are the basic dietary requirements which *C. briggsae* is unable to synthesize.

Blood feeding by hookworms and other nematodes

In the 1880s it was observed during human autopsies that the hookworms *Ancylostoma duodenale* removed from the intestine "leaked" blood from the mouth and anus. The large amount of blood that passed through the hookworms was also noted by physicians in the last century. It was also found that many red blood cells passed unchanged through the gut of the hookworm, and that mucus also entered the intestine. Anaemia was noted to be the main symptom of patients with hookworm disease, further indicating the blood-feeding habits of hookworms.

In 1931 Wells made the first lengthy and systematic observations in a classical study of the feeding habits of *Ancylostoma caninum* in anaesthetized dogs.

The blood may be seen to be propelled by the sucking movements of the oesophagus, which consist of the alternate expansion and contraction of its lumen, brought about by the action of the muscles of the oesophageal wall. Sometimes a considerable amount of clear fluid, mucus and clumps of epithelial cells will be sucked in before blood is obtained, and this material may be also clearly seen as it is drawn into the oesophagus and forced into the intestine. The supply of blood obtainable from the tip of the villus does not seem to be adequate, for the worm will soon release its hold and with vigorous boring movements work its head between the villi to a deeper point of attachment. The material taken in may not consist entirely of blood, but soon a copious supply of blood is tapped and the boring and twisting movements cease. As the body of the worm becomes quiet the oesophageal movements become very rapid and forceful. A droplet of blood appears with great suddenness from the anal orifice. The process may continue indefinitely, blood droplets being ejected frequently throughout the day. The blood so ejected does not appear to have undergone digestive changes.

Wells estimated that each adult hookworm ingested 0.8 ml/day. More recent estimates using radio-labelled red cells provide measurements of 15–63mm^3/day. Other techniques using Cr^{51} labelled red blood cells, provided means of 0.2 ml/day/worm for *A. duodenale* and 0.01–0.6ml/day for *Necator americanus*. From these observations it may be estimated that a human patient with 100 worms could lose up to 50 ml of blood to the hookworms each day! The pathology of the host-parasite relationship is very complex, because many of the red blood cells that enter the hookworm pass into the intestinal lumen of the host. Nearly 50% of the iron in the red blood cells can be reabsorbed by the host. The position of the hookworm and the factors that influence the selection of sites can therefore become critical, and malabsorption syndromes can dictate the disease symptoms.

A proteolytic enzyme has been extracted from the pharynx of *Ancylostoma caninum*. This enzyme was specifically inhibited by serum from a dog made immune to *A. caninum* by repeated infection (Thorson, 1956). This was deduced to be direct evidence that the dog had been in contact with this antigen at the tissue level. Thorson was also able to produce some immunity in dogs to *A. caninum* by infecting dogs with an oesophageal extract containing the proteolytic activity. This enzyme is probably an antiprothrombinase which acts as an anticoagulant during feeding. When placed in nutrient media with glucose, they cannot feed until serum is added. The provision of a sensory feeding stimulus is clearly required, and this has proven a major hurdle in the *in vitro* maintenance and culture of many nematode parasites.

The stomach nematode *Haemonchus contortus,* which is also called "The Barber's Pole Nematode", is a major scourge of sheep. The conspicuous red and white spiralling along the body is produced by the ovaries and the blood-filled intestine. Using red blood cells labelled with Fe^{59}, measurements have shown loss of blood from 0.005 to 0.173 ml/worm/day. It is clear that in lambs a heavy infection can be lethal.

Inorganic P^{32} injected intramuscularly into rats rapidly enters *Nippostrongylus brasiliensis,* which is therefore believed to feed on host tissues. This has proved a most useful technique, because P^{32} does not enter *Ascaridia galli,* which is thus thought to feed on host intestinal contents.

The fact that nematodes are reddish does not indicate that they are blood feeders—many have haemoglobin in their own tissues. The bright red colour of live feeding *Nippostrongylus brasiliensis* is not the haemoglobin derived from the host, but that produced by the nematode. It has been shown to have different optical and oxygen loading characteristics from the haemoglobin of the host's blood. Blood feeders tend to accumulate the haem end product from digested blood in the form of haematin or haemosiderin. (This is similar to the deposition of dark haem pigment in the red blood cells infected with malarial parasites.) A large amount of this pigment can represent a relatively small amount of blood feeding.

The lungs of frogs and toads are almost invariably infected with the large dark nematodes *Rhabdias bufonis* and *Rhabdias sphaerocephala.* The earlier specific name of *R. bufonis* was *R. nigrovenosa,* which referred to the "black vein" running along the worms. These nematodes feed exclusively on blood from the capillaries of the lungs (Colam, 1971*b*). No tissue or mucus was found, even at the electron-microscopic level. *R. bufonis* has a very small stoma and, unlike hookworms, it does not "hold onto" or "tear off" pieces of host tissue. The movements of these lung parasites, possibly aided by the secretions of enzymes, cause sufficient localized mechanical damage to rupture the lung tissues. The blood is then ingested, and the digestion products are absorbed. The red blood cells are very densely packed in the intestinal lumen, and it may be that serum is expelled. The digestion is entirely extracellular, and the ultrastructure of the intestinal epithelial cells is indicative of absorption only. The digestion of blood cells is probably accomplished by enzymes secreted from the pharyngeal glands.

The feeding of whipworms

The pharynx of the genera *Trichuris* (the "whipworm") and *Capillaria* is modified into an elongate thinly-tapering series of cells collectively called the "stichosome". This thin "whip" is usually found deeply embedded in the tissues of the host and is difficult to observe, so the evidence for feeding is somewhat indirect. Red blood cells are found in the intestine, and blood is known to enter the gut of *Trichuris trichiura* in man and *T. vulpis* in dogs (Burrows and Lillis, 1964). Read (1968) however stated that there was insufficient evidence to conclude that blood is the only, or even the major, food source.

Careful sectioning of infected host intestinal tissues have shown "tunnels" in the mucosal layers of both humans and baboons. Furthermore, these tunnels were lined with altered epithelial cells. *Trichuris suis*, in the pig, lies very close to the host's mucosa and is covered by a thin layer of mucus. In the mucosa of infected pigs, the worms are sometimes surrounded by degenerating mucosal cells. There is some indication of a feeding mechanism that is associated with an intimate contact between the host and parasite, and is related to induced cytological changes in the host.

A biologically unusual member of the genus is *Capillaria hepatica*, which has a similar morphology to other species with the long stichosome, but it lives as an adult in the liver parenchyma tissues of mice. If trypan blue, colloidal gold, or ferritin are injected into the circulatory system of mice infected with *C. hepatica*, they do not enter the tissues of the nematode. This suggests that the nematode does not feed directly on the circulatory system or the normal tissues of the host. In a detailed study, Wright (1974) produced electron-microscopic evidence that the head of *C. hepatica* was surrounded by large modified liver cells or "hepatocytes". These cells were modified by increasing in size and in nuclear division. Wright interpreted these findings to mean that the nematode was inducing localized cellular changes in the liver to form "food cells", and that their nutrient was derived from these specialized cells.

There appears to be a growing common theme linking the feeding habits of trichurid nematodes. Their unusual pharynges, and the close association of the worms with modified host tissues, suggest a highly specialized feeding mechanism in the group.

The feeding of plant-parasitic nematodes

Plant-parasitic nematodes may be induced to feed on normal or excised roots grown in agar or nutrient media, and can be observed much more readily than can the parasites in the tissues of animals. As a result there is a large body of information regarding the food and feeding habits of these species (comprehensively summarized by Doncaster (1971)).

All phytoparasites have a stylet or spear with which they puncture host cells, and through which they absorb nutrient. The pore of the stylet is very small, usually being less than $0.5\mu m$; no bacteria can enter, and these forms are exclusively fluid feeders.

In most cases the feeding events are similar and may be summarized as follows:

1. Active movement prior to contacting the plant cells (also called "searching" by Doncaster);
2. Contact of the nematode head with the plant;
3. Reduced bodily movement;
4. Head waving and spasmodic stylet thrusting;
5. Cessation of bodily waves;
6. Deliberate stylet thrusting and the adoption of a characteristic body posture;
7. There may be injection of extracorporeal enzymes into host cells;
8. Rapid bulb pulsations of the pharynx and ingestion;
9. End of feeding and movement away.

This pattern can vary from species to species, but is always of this basic form. In a very full and elegant series of studies using microcinematographic techniques, Doncaster has analysed the movements of the pharynx and shown them to be highly integrated and very complex. The methods of coordination are so far obscure, but the pharynx has a "pharyngeal-sympathetic nervous system" and is affected by turgor changes.

In many species, such as *Tylenchorhynchus dubius* (Wyss, 1973) and *Ditylenchus myceliophagus,* an ingestion period is preceded by the injection of pharyngeal secretions into host cells. As soon as the host cell has been perforated, the granular secretions flow down into the stylet, and a clear region develops in the host cell around the tip of the stylet. Ingestion of this partly pre-digested food follows immediately. *D. myceliophagus* and *D. destructor,* which feed on fungal hyphae, force extracorporeal enzymes into the hyphae by localized muscular contractions and bodily turgor changes.

The most advanced genera of plant-parasitic nematodes are considered to be those in which the adult parasite is a sedentary female living on or embedded in the root tissues of their hosts. There are two main genera in this category: *Heterodera* spp. and *Meloidogyne* spp. In these forms, a complex host-parasite relationship develops in which the sedentary parasites induce "syncytia" or "giant cells" on which they feed. The infective second-stage larvae of these genera move through the soil, contact the root of the host, and penetrate into it. Giant cells begin to form within a few hours on sweet potatoes, and in a few days on tomato roots. Substances are probably secreted from the pharyngeal glands of *Meloidogyne,* which act either as an auxin, inhibiting indole acetic acid oxidase, or they induce the plant to produce an excess of its own growth substances. It is known that 2, 4-dichlorophenoxyacetic acid (2,4-D) inhibits the normal action of natural plant-growth regulators, and it may be that *Meloidogyne* secretes a comparable compound. Syncytia may form through the rupture of host-cell walls, but the induction of nuclear and cytoplasmic changes are much more complex than just mechanical damage. The nematodes are required to maintain the giant cells, which may persist for weeks. If they are removed, the syncytia and giant cells do not persist (figure 5.7).

(a)

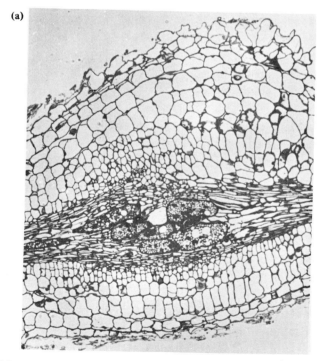

Figure 5.7. Giant-cell formation in tomato roots. Longitudinal sections of tomato roots following invasion by *Meloidogyne incognita* larvae. *(a)* Five days after exposure (×600)

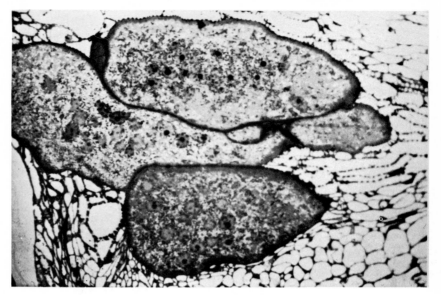

Figure 5.7 *(b)* Three weeks after exposure (×1200) (photographs J. M. Webster).

Esterases, proteases, cellulases, chitinases and pectinases have been found in homogenates of plant-parasitic nematodes. This strongly suggests that they are able to synthesize the enzymes necessary to digest plant tissues.

Bacterial feeding

The simplest form of feeding in nematodes is bacterial feeding. In addition to the oxyurids discussed above (which consume bacteria in the hind guts of their hosts) most soil, marine and freshwater forms are partially or exclusively microbivorous. In culture, the marine chromadoridan *Chromadora macrolaimoides* feeds on 14 species of bacteria, 20 species of algae, and 2 species of planktonic diatom (Tietjen and Lee, 1973).

It is particularly noteworthy, in the context of nematode biology, that many of the parasites of animals such as *Ancylostoma* spp., *Nippostrongylus brasiliensis,* and *Rhabdias bufonis,* which were discussed as blood and tissue feeders, are bacterial feeders in their early larval stages. A number of laboratory studies have shown that the free-living microbivorous larvae of parasites can survive on many species of bacteria, but that some support better growth than others.

Table 5.1. Ingestion of food by *Aphelenchus avenae* feeding on *Botrytis cinerea* (data from de Soyza, 1973).

Day	Volume of Valve (a) μ^3	Mean No. pulses/sec. (b)	Volume of food Ingested/s. in μ^3 $(a \times b) = X$	Feeding Time/15m Period Sec. Y	Vol. Ingested/15 m period XY μ^3	Vol. Ingested for 24 hours $XY \times 4 \times 24$ μ^3	Vol. Ingested per 24 h in ml $\times 10^{-14}$	Cal Ingested Per Day $\times 10^{-4}$
1	34.5496	8.14	281.2337	16.62	4674.1041	448713.9936	44871399	0.3966
2	67.4838	7.47	504.1040	25.99	13101.6630	1257759.6480	125775965	1.1117
3	84.3652	7.50	632.7390	15.66	9908.6927	951234.4992	95123450	0.8407
4	147.8714	6.50	961.1641	10.11	9717.3691	932867.4336	93286743	0.8245
5	200.6264	6.83	1370.2783	54.60	74817.1952	7182450.7392	718245074	6.3484
6	200.6264	6.54	1312.0967	45.15	59241.1660	5687151.9360	568715194	5.0268
7	227.7503	7.21	1642.0797	14.40	23645.9477	2270010.9792	227001098	2.0064
8	227.7503	7.21	1642.0797	23.98	39377.0712	3780198.8352	378019884	3.3412
9	227.7503	7.21	1642.0797	37.33	61298.8352	58864688.1792	588468818	5.2014
10	227.7503	7.21	1642.0797	—	—	—	—	—
16	227.7503	7.21	1642.0797	30.80	50516.0548	4855301.2608	485530126	4.2915
21	227.7503	7.21	1642.0797	17.82	29261.8603	2809138.5888	280913859	2.4829
24	227.7503	7.21	1642.0797	36.47	59886.6467	5749118.0832	574911808	5.0815

$\mu^3 = 10^{-12}$ ml
1 ml = 88.389 cal

Energetics and nutrition

Little attempt has been made to quantify feeding or energy utilization in calorific units. As far as the authors are aware, the study of de Soyza (1973) remains the only investigation of this type. In this study the plant-feeding nematode *Aphelenchus avenae* was cultured in the laboratory, on the fungus *Botrytis cinerea*. As the nematodes feed only on the hyphal contents and not on the fungal cell walls, cytological extracts were made, and the contents dried and combusted in a bomb calorimeter. This enabled an expression of the volume of fungal contents ingested to be made in calorific units.

The nematodes were then observed feeding throughout their life cycles. The daily volume of the ingested fluid was the product of the volume of the

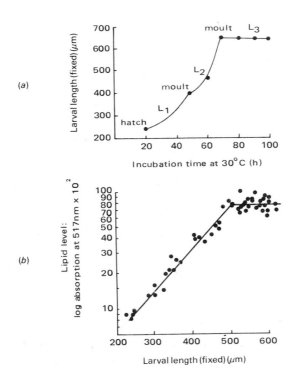

Figure 5.8. *(a)* Growth curve of *Ancylostoma tubaeforme* in faecal culture at 30°C. Moults occur at 50 and 66h. Each point is the mean of 100 larvae. *(b)* The relationship between *A. tubaeforme* larval length and the rate of synthesis of unbound neutral lipid at 30°C. (redrawn from Croll, 1972).

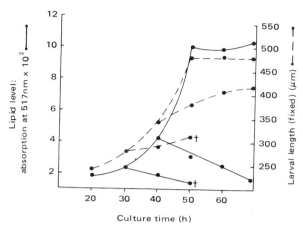

Figure 5.9. The relationship between lipid level and *A. tubaeforme* larval length in faecal culture at 30°C. At 30 to 40h larvae were isolated from bacteria at 30°C, and their length and lipid calculated after incubation. † = death; each point the mean of at least 20 individuals (Croll, 1972).

pharyngeal pump, the rate of its pulsation and the period spent feeding per day (Table 5.1). The data show that the mean pulsation rate remains constant during all stages of the life cycle, in both larvae and adults. The duration of feeding periods varied widely and without a constant pattern. However, the volume of the valve in the pump showed considerable increase in size as the worm grew, and was the major factor accounting for the increased ingestion of calories with growth.

Larvae of the cat hookworm *Ancylostoma tubaeforme* may be readily cultured on bacteria. The increase in volume and density of larvae of *A. tubaeforme* from hatching as a first-stage larva to moulting into a third-stage larva was calculated; this was then expressed in terms of bacterial cells. Assuming a 20% conversion of bacteria to nematode tissue, about 2.5 million bacteria need to be ingested within 2 or 3 days (Croll, 1972).

As larvae of *A. tubaeforme* grow and feed, they synthesize a food reserve of unbound neutral lipid. It is possible to stain the free unbound neutral lipid in Oil Red O, and to measure the absorption of the red colour on a scanning microdensitometer. Using this technique it was found that, as the larvae grew, there was an exponential increase in the amount of lipid being synthesized (figure 5.8). Furthermore, if the larvae were isolated from food during the development of the preinfective larvae, they could continue activity and morphogenesis (figure 5.9) by remobilizing the stored lipid.

CHAPTER SIX

DEVELOPMENT OF NEMATODES

THE SEXES IN MOST NEMATODES ARE SEPARATE BUT, AS WITH ALMOST all generalizations about the group, there are exceptions, and hermaphrodite and parthenogenetic species are known, particularly amongst microbivorous forms.

The male reproductive system may be divided into three sections: the testis, the seminal vesicle and the vas deferens. The commonest arrangement is for the spermatocytes to be produced only at the proximal end of the testis (telogonic) but, in the Trichuroidea and Dioctophymatoidea, the testes are hologonic and spermatocytes are produced throughout their length. Most males are monorchic (with a single testis) but some diorchic males, with paired testes, exist (figure 6.1). The whole system may be very long and coiled, as in *Parascaris equorum* or short and uncoiled, as in many free-living species. The vas deferens opens into the cloaca, which is associated with the male accessory organs, the spicules and, in the Strongylida, the bursa.

In telogonic forms the testis itself may be sub-divided into two parts: a germinal zone at the tip and a growth zone (figure 6.1). The cells may be arranged around, and have cytoplasmic bridges to, a central rachis, as in *Heterakis* (Lee, 1971); or the spermatocytes may just form a row of larger cells. The testis is covered with a thin epithelium which is continuous with the gonoduct. The testis merges into the seminal vesicle, a dilated region in which the maturing sperm are stored. In some large nematodes, e.g. *Ascaris*, there is a vas efferens but, in most, the seminal vesicle passes straight to the vas deferens consisting of an anterior glandular region and a posterior muscular ejaculatory duct.

Almost all male nematodes have at least one, and more usually a pair of copulatory spicules. These vary greatly in shape and size, and are not distorted by fixation (figure 6.1); they consequently have some value to the taxonomist. The spicules are lodged in an invagination of the cloaca; some are hollow with a nerve which is probably sensory running

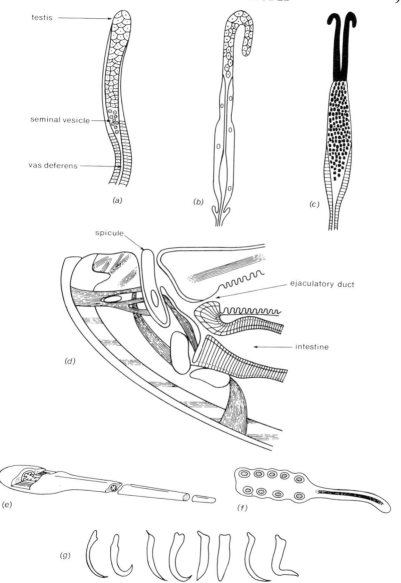

Figure 6.1 Male reproductive system *(a) Chromadora, (b) Rhabditis, (c) Heterodera (d)* section through tail of *Ascaris, (e) Aspiculuris* sperm, *(f) Nippostrongylus* sperm, *(g)* shapes of spicules from different species. (Redrawn from various sources).

through them (see Chapter 3). Each spicule is moved by a pair of retractor muscles, and many nematodes have a cuticular groove or accessory thickenings, the gubernaculum and telamon respectively, which help to guide the spicules. In the Strongylida, the male tail is modified into a bursa which is used to hold the male and female in position during copulation (figure 6.2).

Spermatogonia are produced at the germinal tip of the testis. During a growth phase in the testis they change shape, enlarge and become more clearly defined spermatocytes, which give rise to the spermatids and later mature to spermatozoa in the seminal vesicle. Meiosis occurs when the large, round, late spermatocytes give rise to the spermatids. Foor and McMahon (1973) have shown that, in *Ascaris,* final maturation of the spermatozoa is accomplished with the aid of substances produced by the glandular region of the vas deferens, and is thus probably not normally achieved until after insemination.

Nematode spermatozoa are normally regarded as amoeboid rather than flagellate. Their morphology varies (figure 6.1) but the few that have been examined ultrastructurally all seem to have a large round nucleus with mitochondria and a few strands of rough endoplasmic reticulum. Following meiosis, the chromatin comes together in the centre of the spermatid.

Copulation in nematodes has been observed surprisingly rarely. Some species, e.g. *Syngamus trachea,* the "gape-worm" of chickens, remain in permanent copulation after pairing, but most mate for a short period only.

Females of *Ditylenchus destructor* are receptive to the male for about a week after the final moult. (The males are capable of mating for about three weeks.) Anderson and Darling (1964) have described mating in *D. destructor.* The male approached the female directly; the head first touched the vulva, then twisted away; the male began to move in a spiral around the female with the ventral side in contact. The female tail was bent, which assisted in guiding the male over the vulva. The male generally made several passes over the female before the spicules and vulva were aligned and insemination could commence. Six to twenty spermatozoa were injected within a second into the posterior portion of the uterus, which was held open by the deeply inserted spicules. The male moved away immediately after copulation.

A single female may copulate several times during her receptive period, and with more than one male, although several hours elapse between matings. The fecundity of females is not affected by the number of times mating occurs, so that promiscuity is more likely to be a factor in increasing genetic variation in the progeny than in just ensuring fertilization.

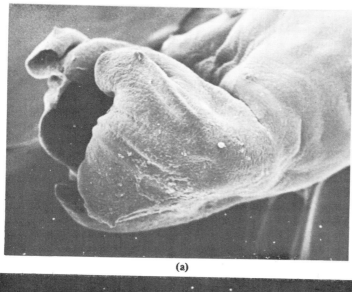

(a)

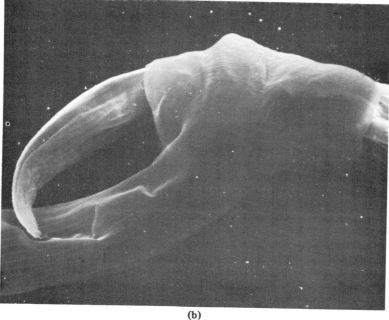

(b)

Figure 6.2. *(a)* Bursa of *Ancylostoma tubaeforme* *(b)* Mating *Nematospiroides dubius* Male bursa folded around the tail of the female.

After insemination, the sperm pass up the female reproductive tract to the spermathecae where they may change from an elongate to a spherical shape. The viability of sperm within the female varies from a few days in some species to three weeks in the dog hookworm *Ancylostoma caninum*.

The female reproductive system is somewhat more complicated than that of the male (figure 6.3). The vulva is slit-like and, by definition, ventral. It opens into a single short cuticle-lined vagina which in some species may act as an ovijector. The vagina normally passes to paired uteri, on each of which is a spermatheca which may just be a dilated area near the junction of uterus and oviduct. The uteri are continuous with paired oviducts, and each leads to an ovary. The position of the vulva varies widely between species, and may be found from the oesophageal to the anal region. This means that the oviducts may be arranged in different positions: where both oviducts are directed anteriorly, the arrangement is said to be *prodelphic*; when both are directed posteriorly, *opisthodelphic*; and when one passes anteriorly and the other posteriorly, *amphidelphic*.

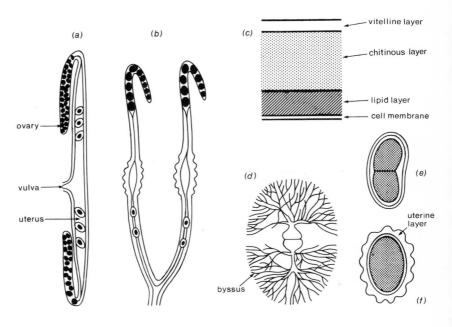

Figure 6.3. Female reproductive system. *(a) Rhabditis, (b) Meloidogyne, (c)* diagrammatic section through egg shell. Egg of *(d) Mermis, (e) Meloidogyne, (f) Ascaris* (Redrawn from various sources)

The ovaries are tubular and may be elongated. Like the testes, the ovaries in the Trichuroidea and Dioctophymatoidea are hologonic with germ cell proliferation along the length of the ovary, but in other families they are telogonic with a limited zone of proliferation near the tip of the ovary. There may be a central non-nucleated rachis to which the developing ova become attached by cytoplasmic bridges.

Oogonia are generally produced at the proximal end of the ovaries and are bounded by a thin membrane. As they pass down the tubular ovary, they increase in size due to accumulated food reserves, normally lipid or glycogen. The distal portion of the ovary is occupied by mature oocytes. The oviducts are narrow tubes that are probably not secretory. Anya (1964) has suggested that they mould the eggs to their normal shape and provide the correct environment for fertilization.

Fertilization occurs within the spermathecae as these are normally full of sperm. In *P. equorum,* entry of a sperm into the egg initiates the formation of a fertilization membrane and the final division of the egg nucleus. These final divisions, one mitotic, the other meiotic, give rise to the two polar bodies and a haploid egg nucleus. The chromosomes of the male and female pronuclei then come together and a mitotic division occurs to give two diploid cells.

Nematode eggs

As biological units, nematode eggs fulfil three major functions: they provide a partially controlled environment for development of the embryo, they offer resistance to external environmental factors and, in the case of parasitic species, they may act as the means of transfer from host to host.

While there are constant and striking differences between the eggs of certain species (figure 6.3), there is nevertheless a basic similarity and, considering the vast difference in size of the adult females, the eggs are of remarkably uniform size, most being within the range 50–90μm $\times$ 20–50μm.

The egg shell consists of three main layers secreted by the egg itself (figure 6.3). The inner is a lipid layer which in *A. lumbricoides* consists of lipoprotein with 25% protein. The composition of this layer varies with species, and it is thought that the differences may reflect differences in the ability to resist desiccation. The middle layer is composed of chitin and protein, and is the only chitin-containing structure that has been recorded in nematodes. The chitin is produced by polymerization of acetylglucosamine and is synthesized after fertilization. The amount of chitin differs between species; eggs of *Heterodera* contain 9% chitin and 59% protein, whereas in *Ascaris* the layer is largely chitin and there is very little protein. Outside the chitinous layer is an outer lipoprotein vitelline layer. This latter layer is the one that bounds the egg prior to fertilization, and the rest of the shell is laid down within it. In some species, e.g. *Ascaris* and *Mermis*, there is a fourth uterine layer secreted by the uterus which consists of a gelatinous matrix and an acid-mucopolysaccharide tanned-protein complex. This uterine layer is the one that is often characteristically sculptured.

The egg production by females varies widely (Table 6.1). Large parasitic species produce prodigious numbers. Adult female *Ascaris* have been found to contain an estimated 27,000,000 eggs and may lay 1,500,000,000 in a lifetime. Free-living and plant-parasitic species tend to produce fewer eggs, and not all animal parasites produce the large numbers of *Ascaris*.

Embryology

The horse ascarid *Parascaris equorum* (originally known as *Ascaris*

Table 6.1. Fecundity of various nematode species

ANIMAL PARASITES

Ascaris lumbricoides	200,000 eggs/female/day
Necator americanus	10,000 " " "
Strongyloides stercoralis	50 " " "
Dracunculus medinensis	1,500,000 larvae/female

PLANT PARASITES

Heterodera rostochiensis	200 eggs/female
Anguina agrostis	1,000 " "
Meloidogyne javanica	350 " "

FREE LIVING

Caenorhabditis elegans	240 eggs/female
Pelodera teres	360 " "
Prionculus punctatus	45 " "

megalocephala) may cause diarrhoea and intestinal disturbance in foals, but it is not of great significance as a pathogen. However, this parasite has played a very important role in the development of our understanding of nematode embryology and, perhaps of greater significance, in fundamental aspects of cytology and heredity.

Following the enunciation of the cell theory by Schlieden and Schwann in 1839–40, there was a great upsurge of interest in the cell. During the period 1840–70, the essentials of the cell theory were delineated and the concept of genetic continuity defined. This period was followed by one which included the development of cytology and cellular embryology, and which gave form to the general ideas on the physical basis of heredity and the mechanism of development. In the latter part of the nineteenth century the eggs of *P. equorum* were used extensively. Von Beneden (1883) while studying the history of the nuclei during fertilization, first demonstrated that both parental cells contribute an equal number of chromosomes to the nuclei of the offspring. Hertwig (1890) followed this by establishing the

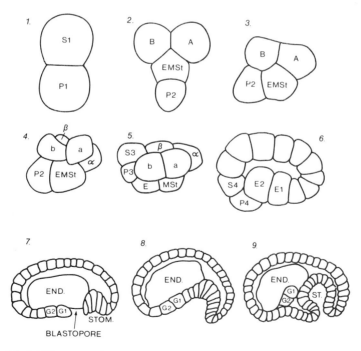

Figure 6.4. *(a)* Diagrammatic representation of nematode embryology.

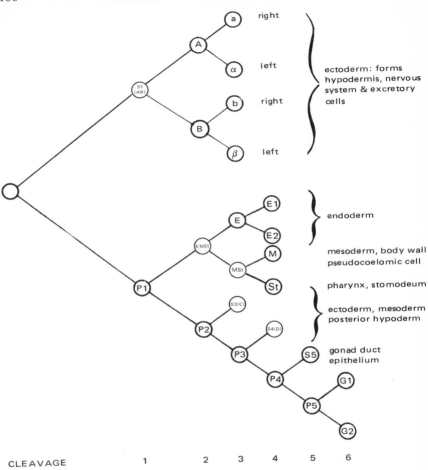

Figure 6.4. *(b)* Summary of nematode cleavage. See text for explanation.

nature of the polar bodies. Then Boveri (1899) illustrated the determinate form of development, and also the possibility of the cytoplasm being partially responsible for controlling development.

The embryonic development of nematodes has been observed in relatively few species, but the pattern of development in each case examined has been very similar, although differences in detail have appeared. The work of Zur Strassen (1896) and Boveri (1899) is probably the most detailed, and established the nomenclature that has been used by subsequent workers. Essentially, nematode embryology may be described as displaying determinate cleavage with subsequent mosaic development; although highly determinate, nematode cleavage does not follow the typical spiral pattern, and nematodes are almost alone among animals in that the first cleavage is equatorial, cutting the egg axis horizontally; this corresponds roughly to the horizontal plane of the embryo.

The first cleavage results in two blastomeres, an anterior, slightly larger one, designated S_1 (or AB), and a posterior one P_1 (figure 6.4a–1). The nucleus of the S_1 cell undergoes the extraordinary phenomenon of chromatin diminution, in which the chromosomes fragment, and much of the chromatin material is absorbed. The remaining fragments each act like a minute chromosome at the next division. It is thus possible to distinguish the somatic from the parental cells at this early stage. S_1 divides to give two cells A and B, and P_1 divides into EMSt and P_2. EMSt undergoes chromatin diminution, leaving P_2 as the only cell that retains the total chromosome complement.

In *P. equorum*, S_1 divides longitudinally and P_1 transversely to give a T-shaped embryo, and the cells rearrange to give a rhomboid with A and B anterior and dorsal, EMSt ventral and P_2 posterior (figure 6.4a–3). In other species, S_1 and P_1 both divide transversely to give a linear arrangement, and reorientation occurs at a later stage. A and B and their descendants comprise a dorsal cell group, and EMSt a ventral cell group; cleavage in these two groups is not necessarily synchronous.

At the third cleavage, A and B divide into a and α, b and β respectively, with a anterior and b posterior; a and b form the right, and α and β the left side of the dorsal ectoderm. The bilateral symmetry of the embryo is thus established at this early stage in development. As cleavage continues, the dorsal cells increase in number and eventually form the anterior hypodermis, nervous system and the excretory cells. EMSt divides into E, which gives rise to the endoderm, and MSt which at the fourth cleavage forms M, the stem cell of the mesoderm, and St, which divides to give rise to the pharynx and stomodeal structures.

The parental germinal cell P_2 divides to give S_3 (C) and P_3; the latter divides again to give P_4 and S_4 (D); S_3 and S_4 undergo chromatin diminution and then continue division, forming ectodermal and mesodermal tissues, and also the cells of the posterior hypodermis. Division of P_4 gives P_5 and S_5, which gives rise to the gonad duct epithelium. By this stage all the cell types have been differentiated, and only a single cell P_5 retains the full chromosome complement. This cell divides only once more, to give

G_1 and G_2, until the larval stages are completed and the genital organs are formed.

The embryo is now in the form of a hollow ball of cells, the blastula, and gastrulation ensues. Gastrulation may be by epiboly and overgrowth of the ectoblasts, or probably more typically by invagination, where there is an increase in the number of ectoblasts, and the mesoblasts are pushed into the blastocoel; these are followed by the germinal cells, G_1 and G_2, and the blastopore closes as the lateral lips fuse. Gastrulation results in the formation of a cylindrical embryo. Cell division continues, and the embryo elongates and becomes coiled within the egg.

Nematode embryology exhibits a number of interesting features. The process of chromatin diminution means that only in the germ cell line is the full chromosome complement retained, and the germ cell line is traceable directly from generation to generation. The highly determinate nature of the mosaic development is probably better demonstrated by the nematodes than by any other group of animals. Destruction of one of the blastomeres at even the two-cell stage results in the development of the remaining cell as it would in the normal embryo. Destruction of any of the later cells results in an embryo deficient in the cells derived from the missing cell.

Although the process of development described may appear a relatively static one, it is highly dynamic. The cytoplasm of the fertilized cell prior to the first cleavage may show vigorous movement, and there is

Table 6.2. Time required for full embryonation of various nematode species.

Habit	Species	Temperature °C	Time to hatching or full embryonation
Free living	*Rhabditis teres*	18	20h
	Acroboles complexus	25	4–5 d
Plant Parasitic	*Pratylenchus penetrans*	23	10 d
	Hemicriconemoides chitwoodi		15–19 d
	Xiphinema spp.	20–25	19–24 d
	Longidorus spp.	20–25	25–30 d
	Meloidogyne nassi	22–26	15–17 d
Animal Parasitic	Trichostrongyle spp.	20–23	24 h
	Ancylostoma caninum		24 h
	Necator americanus	30	20–30 h
	Ascaris lumbricoides	30	18–20 d
	Parascaris equorum		10 d
	Trichuris trichiura	30	21 d

continual cellular movement throughout development. Co-ordinated movement of the whole embryo is initiated shortly after gastrulation, and continues more or less continuously until the egg hatches.

The time from fertilization to hatching varies widely between species (Table 6.2) probably because, while most nematodes hatch as first-stage larvae, such diverse species as *Heterodera rostochiensis* and *A. lumbricoides* undergo one moult within the egg and hatch as second-stage larvae. In general, those nematodes that hatch to give active free-living larvae tend to hatch more rapidly than plant-parasitic species, or those animal parasites that rely on ingestion of the embryonated egg for transmission. In the latter cases, a stimulus from the host may be required before hatching is initiated. The temperature may have a great effect on the time required prior to hatching. For two species of Diplogasterinae, Pillai and Taylor (1968) found that hatching would occur between 10° and 35°C, but the time required varied from 85h at 10°C to 7.5h at 35°C (figure 6.5). Crofton (1965) obtained similar results for a range of pasture nematodes parasitic in sheep. Depending on the individual species, hatching occurred between 38°C and 4°C, requiring from 11h to 7d to do so.

Hatching

The emergence of nematode larvae from their eggs has been studied in relatively few species, and it is not easy to generalize about the processes

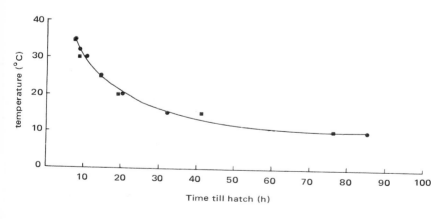

Figure 6.5. Influence of temperature on the time required prior to hatching of ●———●*Fictor anchicoprophaga* and ■———■*Paroigalaimella bernensis* (data from Pillai and Taylor, 1968).

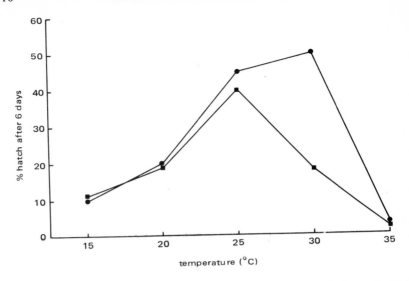

Figure 6.6. Influence of temperature on the hatching of ■————■ *Meloidogyne hapla* and ●————● *M. javanica* (redrawn from Bird and Wallace, 1965).

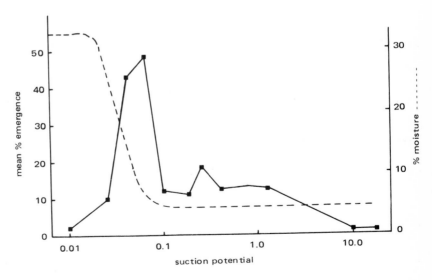

Figure 6.7. Percentage larval emergence from *Meloidogyne javanica* eggs at a range of osmotic potentials (redrawn from Wallace, 1966).

involved. Those species that have been examined do, however, fall into two general categories: those nematodes which hatch "spontaneously" when the larva is fully developed, and those which require more or less highly specific stimuli provided by a potential host before eclosion. This latter "stimulated" hatch has also been called "infectious" hatch (Rogers, 1962).

"Spontaneous" hatching occurs in free-living nematodes, and many of the plant and animal-parasitic species which have a free-living phase in their life cycle. The term "spontaneous" does not imply that hatching is a wholly intrinsic phenomenon and that environmental factors do not play a part. Temperature, moisture-stress and the composition of the gas phase, all influence emergence (figures 6.6, 6.7). "Spontaneous" hatching eggs tend to be less resistant to extremes of environmental conditions than those which require host stimuli. However, there are exceptions to this generalization, e.g. the eggs of the sheep intestinal parasite *Nematodirus battus* require a period of cold, for hatching to occur when the temperature again rises. This modification of the hatching process has survival advantage, as most adult worms do not survive the winter in the sheep intestine; eggs laid on to the pasture do not hatch until next spring, when the next generation of susceptible lambs is available for infection.

Eclosion in those "spontaneously" hatching species that have been studied follows a roughly similar pattern. Movement of the larva within the egg becomes more continuous and better co-ordinated, and oesophageal pumping and activity in the sub-ventral oesophageal glands have been reported in some species (Bird, 1968; Croll, 1974).

For the intestinal parasite *Trichostrongylus retortaeformis* which hatches spontaneously, Wilson (1958) found that a combination of movement within the egg and release of an enzyme broke down the lipid layer; the shell became permeable and water entered. This increased the volume of the egg contents and pressure on the egg shell, which then ruptured. Although no chitinase was demonstrated, it has been postulated that the larvae may produce such an enzyme to weaken the egg shell and assist rupture, as occurs in *Ascaris*.

If this hatching mechanism is correct, changing the osmotic pressure of the medium from which the eggs hatch would be expected to influence the uptake of water through the permeable egg-shell, and hence the volume increase and the pressure exerted on the shell. Croll (1974) showed just such an effect with a related strongyline, the human hookworm *Necator americanus*. Embryonation was not affected by osmotic stress, which suggested that the integrity of the egg shell was retained, but the percentage hatching decreased with the increase in osmotic pressure (Table 6.3). Influx

Table 6.3. The influence of different salinities on the development of *Necator americanus* larvae in the egg, and upon emergence at $30°C \pm 4°C$ (Croll, 1974).

Incubation Solution (% NaCl)	% Embryonated after 20h	% Hatched after 40h
Distilled water	92	36
0.1	90	56
0.2	79	46
0.4	93	38
0.8	86	48
1.0	93	11
2.0	78	0
4.0	90	0

of water was demonstrated by a marked increase in egg volume prior to emergence.

Specific evidence for enzyme activity during the hatching process has not been demonstrated, but such activity has been postulated on many occasions, and changes in the plasticity of the egg shell occur in a number of species, e.g. *Tylenchorhynchus maximus* (Bridge, 1974); *Meloidogyne javanica* (Bird, 1968). Spontaneous hatching in these plant-parasitic species tends to be a longer process than in animal parasites. Softening of the egg shell in *T. maximus* began 12h prior to hatching, whereas in *N. americanus* behavioural changes are only seen 90min before eclosion.

Stylet-bearing nematodes may make use of the stylet to initiate rupture of the egg shell. Doncaster and Shepherd (1967) showed that *H. rostochiensis* could systematically cut its way out of the rigid egg-shell, but most of the other species examined appear to use the stylet only to puncture the already stretched wall of the softened egg, and thus initiate the formation of the slit through which emergence occurs. Emergence is generally head first and, in some cases, if the split in the egg's shell occurs away from the anterior end, the larva is unable to escape from the shell. However, in other species, tail-first emergence is a not-uncommon occurrence.

The factors responsible for the initiation of hatch of eggs which require extrinsic stimuli from their potential hosts have received considerable attention, as many of the species concerned are parasites of man, his animals and crops. In these species, the infective agent may be the egg itself rather than a motile larva, and a more or less specific stimulus for hatching ensures that eggs do not hatch in isolation from their host. Eggs may remain for long periods until a suitable host appears and, in general, the eggs of these species are more resistant to environmental extremes.

"Infective" hatching has been most fully studied in the animal parasites *Ascaris, Ascaridia* and *Toxocara,* and in the plant-parasitic *Heterodera* spp.

When fully embryonated *Ascaris* eggs containing second-stage larvae are ingested by a host, they pass down the gut to the small intestine where they begin to hatch. Experiments implanting eggs directly in the intestine have shown that treatment in pre-duodenal host gut fluids is not required to initiate hatching, nor is the host directly responsible for digesting the egg shell. Exposure to a suitable set of stimuli initiates the formation in the egg of a "hatching fluid" which can continue to work even if the stimulus is removed. Rogers (1958, 1960) found that *in vitro* at 37° C the presence of CO_2 was required to initiate hatching. A reducing environment and a pH of about 8 increased the percentage hatching, but only if CO_2 was present.

The developing larvae of *Ascaris* are bathed in a fluid containing trehalose. Once hatching has been stimulated, the trehalose level in the medium containing the eggs increases (figure 6.8), illustrating an increase in the permeability of the shell. There is also an increase in the level of N-acetyl-glucosamine in the medium (figure 6.9), produced from the breakdown of the chitin in the shell by a chitinase produced by the larva. When it finally emerges from the egg, the larva is still surrounded by the lipid layer which ruptures or is digested to allow it to escape. The fact that the normally impermeable lipid layer is retained after the egg shell has become permeable, suggests that the first action of the hatching stimulus may be to cause a change in this layer. The dialysed supernatant removed from hatched *Ascaris* eggs has been shown to have lipase and chitinase activity, and it has been suggested that a protease may be present (Rogers, 1958). This fluid has the capacity to initiate hatching in unstimulated eggs.

Heterodera spp. parasitize and damage the roots of a wide range of host plants. The females develop within the root but, as they mature, they break through the surface and lie almost completely in the soil, with only their heads embedded. Eggs are retained within the uterus and, as the females mature, the cuticle, which is normally whitish in colour, is transformed into a brown protective cyst. The eggs embryonate, and the first larval moult occurs within the eggs in the cyst. Hatching occurs in response to external

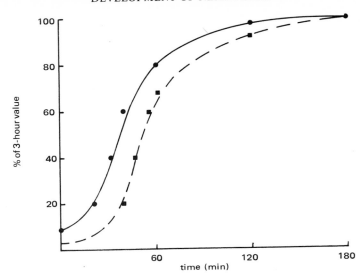

Figure 6.8. Leakage of trehalose from *Ascaris lumbricoides* eggs following stimulation to hatch. Figures represent the percentage of the 3-hour result.
● ────── ● Trehalose leakage ■ ──────■ Hatching (redrawn from Fairbairn, 1960).

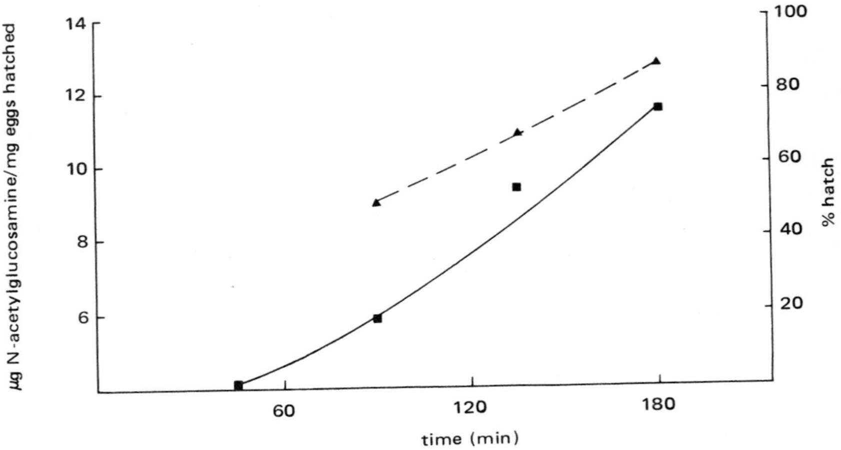

Figure 6.9. Release of N-acetyl-glucosamine into the medium by *Ascaris lumbricoides* eggs following stimulation to hatch ■────── ■ percentage hatching ▲─────▲ (data from Rogers, 1962).

stimuli provided by substances produced by the roots of potential host plants; these are collectively known as "root diffusates".

The economic importance of the plants parasitized by *Heterodera* has resulted in a great deal of work in an attempt to control the nematodes. One approach has been to attempt to hatch the eggs and release the short-lived larvae into the soil well before planting susceptible crops. This control method has not proved very successful, but it has resulted in a wide range of chemicals being screened for their activity in inducing hatching. The ease with which different species hatch *in vitro* appears to reflect the host range of the species concerned (Table 6.4). Considerable effort has been expended in an attempt to characterize the natural hatching factor for *H. rostochiensis*, which affects both tomatoes and potatoes. It is generally accepted that a single unstable organic-acid factor is involved, but final chemical isolation and identification from root diffusate has not been achieved.

Moulting and exsheathment

As far as is known, all nematodes have six stages in their life cycle: egg, four larval stages, and adult. The last five stages are separated by cuticular moults.

The importance of the cuticle in establishing many of the unique features of nematodes has been emphasized earlier, and it is perhaps not surprising, in view of the different habits of various species, that differences are shown in cuticular structure. The number of species that have been examined

Table 6.4. Influence of hatching agents on eggs of *Heterodera* spp. (from Shepherd (1962); Clarke and Shepherd (1966, 1968).

Species	Host family	Water hatch *in vitro*	Artificial hatching agents	Hatching response in root diffusates.
H. avenae	Graminae	Up to 15%	None	None
H. rostochiensis	Solanaceae	1%	10 inorganic ions Anhydrotetronic acid Picrolonic acid	Active to many host spp.
H. schachtii	Cruciferae Caryophyllaceae Umbelliferae Labiatae Cheropodiaceae Amarantaceae Polygonaceae	Up to 25%	21 inorganic ions Wide range of organic compounds	Active to many host spp.

ultrastructurally is limited, and differences in detail occur. The larger adult nematodes, in general, have a more complex structure than small adult or larval stages.

Essentially the cuticle is composed of three main layers: an outer cortical layer, a central median layer, and an inner basal layer (Bird, 1971). Each of these layers may be composed of several constituent layers, depending on the species. Within the basal layer is the hypodermis, and it is a thickening of the hypodermal tissue that heralds the start of the moult. The new cuticle is laid down, starting with the outer cortical layers and working inwards, between the junction of the hypodermis and the basal layer of the old cuticle (figure 6.10). The old and new cuticles then separate, and resorption of some of the constituents of the old cuticle may occur. The cuticular linings of the oesophagus, vulva and intestine, and cuticular structures such as the stylet, are replaced at each moult.

It is not certain what factors govern the initiation of moulting. Neurosecretory granules have been shown to be associated with secretory cells in *Phocanema decipiens* at the time of moulting (Davey and Sommerville, 1974), and it has been suggested that moulting in nematodes may be controlled by a similar mechanism to ecdysis in insects. Insect juvenile hormone and its mimic, farnesyl methyl ether (FME), have been used to initiate normal moulting (Davey, 1971), but whether these

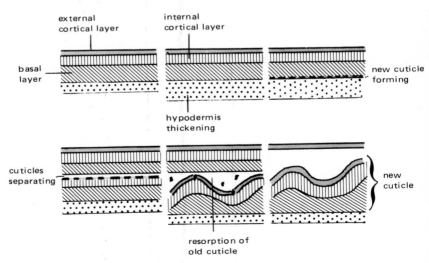

Figure 6.10. Diagrammatic representation of the stages in nematode moulting (after Bird and Rogers, 1965).

hormones are actually involved within nematodes has still to be resolved.

Once the new cuticle has been formed, the moulting nematode is still surrounded by the remainder of the previous-stage cuticle that may or may not have been partially resorbed and from which it must escape. For most moults this is the spontaneous culmination of the moulting process but, in certain cases, this stage is delayed and the cuticle is retained as a sheath. Strictly, the old cuticle at any moult may be regarded as a "sheath" and the process of escape as exsheathment; however, these terms are normally reserved for those specialized occasions when the cuticle is retained and only discarded in response to an exogenous change. This occurs typically in the free-living infective larvae of animal-parasitic strongyle and trichostrongyle nematodes. The exact function of the sheath is not clear, but there is evidence that ensheathed larvae are better able to resist desiccation (Ellenby, 1968), which may increase their ability to survive under field conditions.

The infective stage of many intestinal strongyle parasites is the third-stage larva within the second-stage cuticle, the sheath. Exsheathment occurs only after the larvae have been ingested by a suitable host. Lapage (1935) showed that exsheathment took place in three stages (figure 6.11). The first stage was marked by the formation of a refractile ring about 20 μm from the anterior, after which the cuticle separated into two layers at this ring; generally the anterior broke off as a cap, and the larva escaped. If the weakened cuticle did not fracture, the whole anterior 60 μm of the sheath became distended and further split into two layers, until movements of the larva caused it to rupture.

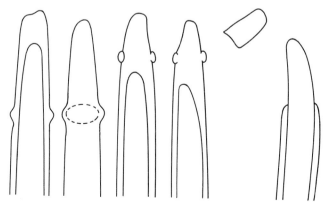

Figure 6.11. Stages in exsheathment of infective trichostrongyle larvae (after Lapage, 1935).

Sommerville (1954) showed that if ensheathed *Haemonchus contortus* larvae were enclosed in a cellophane dialysis bag and inserted into the rumen of a sheep, they exsheathed in 2h. As host proteolytic enzymes were unable to enter the bag, it was concluded that exsheathment was due to endogenous processes. When placed in rumen fluid removed from fistulate sheep, the *inner* surface of the sheath appeared to be digested; if the larvae were surgically removed from their sheaths and the sheaths incubated with rumen fluid, no digestion occurred. However, if large numbers of worms were allowed to exsheath in a cellophane sac, the fluid removed from the sac caused empty sheaths to undergo a typical exsheathment pattern. These and similar results from other species led Rogers and Sommerville (1957, 1960) to suggest that the host gut provided a stimulus for the larvae to produce an exsheathing fluid which was directly responsible for refractile ring formation and subsequent exsheathment. By systematically ligaturing larvae with nylon thread, the area responsible for producing the exsheathing fluid was delineated as being 70–130μm from the anterior and around the excretory pore and the base of the oesophagus. Ultra-violet radiation using a 60–μm beam showed normal exsheathment in all worms except those irradiated in this area. When infective larvae were incubated with rabbit anti-exsheathing fluid antiserum, a precipitate was formed around the excretory pore. The conclusion drawn was that the exsheathing fluid was released from the excretory pore and produced in the region between the excretory pore and the base of the oesophagus.

Table 6.5. Relation between exsheathment *in vivo* and *in vitro*, and the adult site of sheep intestinal trichostrongyle nematodes (Sommerville, 1957).

Species	Exsheath-ment site	pH	Adult site	In vitro exsheathment criteria	
				CO$_2$	pH
H. contortus	Rumen	Neutral	Abomasum	High	Neutral
T. axei	Rumen	Neutral	Abomasum	Low	Neutral
O. circumcincta	Rumen	Neutral	Abomasum		
T. colubriformis	Abomasum	Acid/Pepsin	Small intestine	Low	Acid
Nematodirus spp.	Abomasum	Acid	Small intestine		Acid/HCl
Oesophagostomum colubrianum	Small intestine		Large intestine		

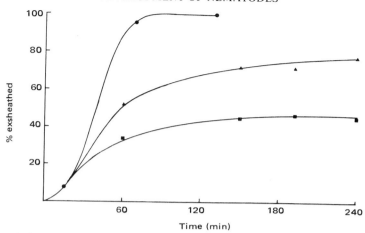

Figure 6.12. Exsheathment of *Ostertagia circumcincta*. ●————● incubated in rumen fluid throughout. ▲————▲ incubated in rumen fluid 15 minutes, then water at 38°C. ■————■ incubated in rumen fluid 15 minutes, then water at 14°C. (redrawn from Rogers and Sommerville, 1960).

In vitro the exsheathing stimulus for trichostrongyle larvae can be provided by gaseous carbon dioxide or undissociated carbonic acid at a temperature of about 38°C, a reducing environment and an appropriate pH enhanced exsheathment (Table 6.5). These results appear to relate to the conditions required *in vivo*, and trichostrongyle larvae tend to exsheath further anterior than the adult site.

Once initiated, exsheathment continues, even if the stimulus is removed (figure 6.12). This has been interpreted as evidence for a "trigger" mechanism which, once activated, continues to release exsheathing fluid, even if the activating stimulus has been removed.

Recently there has been considerable controversy concerning the chemical nature of *H. contortus* exsheathing fluid. This controversy has arisen from the claims of Rogers (1964, 1965, 1970) and counter claims of Ozerol and Silverman (1969, 1970, 1972) over the role of leucine amino-peptidase (LAP) as the active factor in exsheathing fluid. Ozerol and Silverman (1972) have reviewed the evidence for and against LAP and, having concluded that it was not responsible for exsheathment, decided that until further purification of exsheathing fluid is achieved it is not possible to give specific names to the hydrolases which may be the active ingredients. To further complicate the issue, Whitlock (1971) has re-proposed an alternative hypothesis for exsheathment, and the whole question has still to be resolved.

CHAPTER SEVEN

PATTERNS OF NEMATODE LIFE-CYCLE

TYPICALLY THE NEMATODE LIFE-CYCLE CONSISTS OF SIX STAGES: AN egg, four larval, or more correctly, because there is no metamorphosis, juvenile stages, and the adult (figure 7.1). In free-living forms this cycle is completed with little or no modification in a single environment, be it soil, fresh water or the sea. Plant-parasitic nematodes show little specialization in their life cycle beyond a capacity to enter an anabiotic state until suitable host plants are available. By their very nature, plants are static, and the onus is thus on the parasite to seek its host. The adaptations shown by plant parasites are directed towards that end, the non-specific attraction to CO_2 (Klingler, 1970) illustrating such an adaptation. The mobility of animal hosts has resulted in animal parasites that have evolved complex life cycles and behaviour patterns to

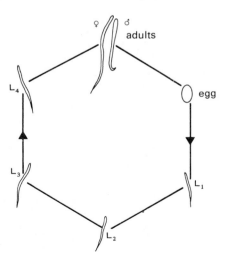

Figure 7.1. Diagrammatic representation of a typical nematode life-cycle.

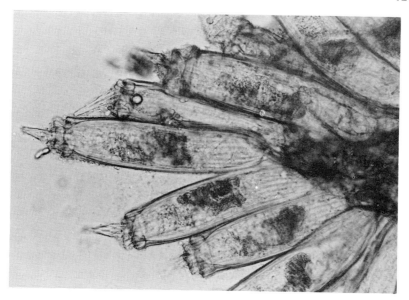

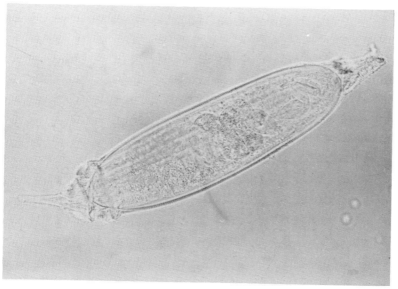

Figure 7.2. *Pelodera coarctata (a)* Group of larvae attached to limb of dung beetle *Aphodius.*
(b) Single ensheathed third-stage larva.

achieve contact. It is thus amongst the nematodes associated with animals that life cycle modifications are most marked.

Individual specific variations in life cycle are legion, and space precludes a comprehensive review. Certain trends occur among animal-parasitic nematodes that illustrate possible paths taken from a free-living to a parasitic way of life, and an increasing dependence on the host.

Of the various levels of more or less permanent association between animals, the least exacting for either party is probably the phoretic one, where one species, the host, acts as a passive carrier of the resistant stage of the second species. *Pelodera coarctata* is a rhabditid nematode that is commonly found in dung, where it continues a normal free-living life-cycle. When conditions in the dung deteriorate, a special resistant ensheathed third-stage "dauer" larva is produced. This larva attaches to dung beetles of the genus *Aphodius*. Beetles may become covered with larvae, which have a characteristic shape due to shrinkage of the sheath (figure 7.2). The larvae remain in this dormant immobile state until the beetle migrates to fresh dung, when they emerge and initiate a further population of free-living nematodes.

Another species of the same genus *P. strongyloides* also shows a phoretic relationship, this time with murine rodents. Third-stage larvae and pre-adults are commonly found within the lachrymal fluid of voles, especially following periods of drought. It is possible that the phoretic stages are produced by the free-living population in response to dehydration or other environmental changes to allow transfer and perpetuation of the species in favourable conditions. The factors governing the change from free-living to phoretic stage and *vice versa* have been little studied, and many details of the processes involved are still unclear. "Phoresis", while it may show dependence, cannot be considered to fulfil any of the criteria of parasitism.

The step from a phoretic life-cycle, such as that of *P. strongyloides,* to a truly parasitic one requires that the transported stage migrates deeper into the host tissue and survives and develops there. A few nematodes have "double" life-cycles with both free-living and parasitic phases, e.g. *Rhabdias bufonis* in amphibians and *Strongyloides* spp. in mammals (figure 7.3). Free-living adults produce eggs that hatch in the soil to give rise to free-living larval stages. These may develop into free-living adults to complete the cycle or, under the influence of factors that are again not fully understood, infective third-stage larvae may develop. These do not develop further until they have entered a suitable host. Entry is normally achieved by penetration through the integument, and larvae migrate through the circulation to the lungs where, in the case of *R. bufonis,* the adults develop.

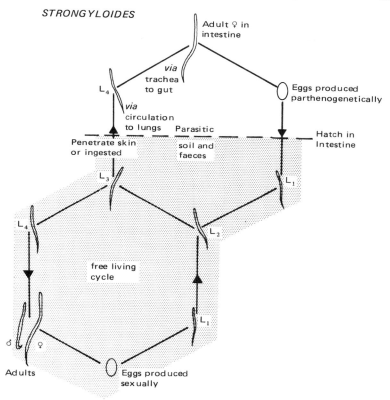

STRONGYLOIDES

Figure 7.3. Diagrammatic life-cycle of *Strongyloides*.

In *Strongyloides* the pre-adult (L_4) larvae migrate to the intestine where the final moult occurs. Parasitic males do not occur, and the adults are protandrous hermaphrodites. The eggs produced hatch in the intestine, and larval stages are released in the faeces to produce either free-living or infective third-stage larvae.

In man *S. stercoralis* is able on occasion to infect the same individual without leaving the host. Infective larvae are produced in the faecal material in the rectum, and may migrate through the peri-anal skin and re-institute the parasitic cycle. Patients in England still occasionally show symptoms of strongyloidiasis, although they have not been subject to reinfection from new infective larvae since the Second World War, thirty years ago.

The nematode parasites of man

The success of any interspecific relationship depends on the whole biology of both partners, so, to eliminate variation in the host and to emphasize the nematode parasites, it is intended in the first part of this chapter to concentrate on the nematode parasites of man. There are two major reasons for this: firstly they are as well studied as any parasites and, secondly, as Heyneman (1966) has suggested, they may present an evolutionary sequence. Initially it must be emphasized that the nematodes discussed are *not* closely related phylogenetically, and those selected for discussion have been chosen because they fit into a pattern; there is little or no evidence that the sequence cited has in fact been followed by any species.

Infective third-stage hookworm larvae penetrate the skin of their hosts, although ingestion of infective larvae also results in infection. Migration via the circulation to the lungs and thence to the trachea follows, and the developing adults pass through the oesophagus and stomach to the small intestine (figure 7.4).

In the intestine both male and female adults occur; copulation takes place, and the females lay eggs that pass out with the faeces. The eggs hatch and first-, second- and infective ensheathed third-stage larvae are produced in the faeces. Infective larvae migrate from the faeces and penetrate the skin of any suitable host that they contact. Larvae may nictate to increase the chance of locating a host. This behaviour occurs only under rather limited

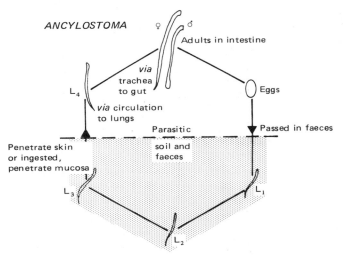

Figure 7.4. Diagrammatic life-cycle of *Ancylostoma*.

conditions of humidity, and may be a response to the temperature and expired carbon dioxide from a potential host (see p. 66).

The first- and second-stage larvae are microbivorous with a typical rhabditiform oesophagus; the oesophagus changes at the second moult, and the exsheathed larvae do not feed but live on stored lipid reserves. The free-living stages of most helminth parasites, e.g. miracidia and cercariae of digeneans, coracidia of cestodes and onchomiracidia of monogeneans, do not feed and have a relatively short free-living existence, a matter of hours as opposed to the days or weeks that hookworms may spend out of a host. Thus, although hookworms do not have a full free-living cycle and are obligate parasites, they do retain a free-living feeding and dispersive stage to bridge the gap between hosts.

The infective larvae of hookworms and other strongyle nematodes are susceptible to environmental changes and, depending on the species, are more or less resistant to extremes of temperature, moisture and tonicity. These factors influence the geographical range of the parasite; many species are unable to develop at low temperatures, or to withstand the winter in temperate climes, and are thus restricted to more tropical areas. Highly-resistant infective stages are produced by some nematodes which withstand climatic variations; these stages are normally non-motile eggs

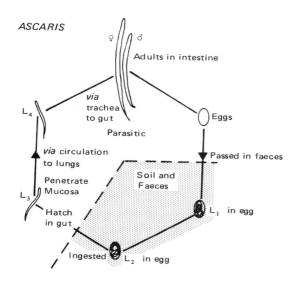

Figure 7.5. Diagrammatic life-cycle of *Ascaris*.

rather than the motile larvae produced by strongyle nematodes.

The eggs of *Ascaris lumbricoides* are notoriously resistant and long-lived; they are produced by adults living within the intestine (figure 7.5) and are unembryonated when passed in the faeces. Embryonation occurs in the soil in about three weeks under optimum conditions; the first moult occurs within the egg so that the fully infective eggs contain second-stage larvae. These eggs may remain without development or deterioration in moist soil for years and only develop further if ingested by a suitable host. Hatching occurs in the gut and the emerging larvae undergo a migration round the body. They penetrate the gut mucosa and pass via the circulation to the lungs, thence up the trachea and back through the oesophagus and stomach to the intestine. During the migration, two moults occur, so that it is the pre-adults that arrive in the intestine where the final moult to the adults occurs. It should perhaps be noted here that Sprent (1962) in a series of studies of the life cycle of the Ascaridoidea, has suggested that the migration seen in *Ascaris* is a reflection of the fact that it has evolved from species with more than one host in their life cycle.

The life cycle of *Trichuris* (figure 7.6) is superficially similar to that of *Ascaris*. However, there is no migration·within the host, and all the parasitic stages take place in the caecum and large intestine. After hatching in the region of the caecum, the larvae enter the villi and moult three times to give rise to the adults. The adults live with the anterior buried in the mucosa (Jenkins, 1970), mate, and the eggs are released in the faeces. The cycle is completed when fully embryonated eggs containing second-stage larvae are ingested.

The infective egg of the human pinworm *Enterobius vermicularis* contains the first larval stage (figure 7.7). Eggs are released by the gravid females when they migrate from the anus onto the peri-anal skin or are passed in the faeces. Contact with the air causes the females to deposit their embryonated eggs which are infective within a few hours. The migrations of the females cause itching which, especially in children, results in scratching and the transfer of eggs from the anal region to the mouth on the fingers. The eggs are easily dispersed in the air and upon ingestion hatch in the intestine. The larvae temporarily burrow into the mucosa in the region of the caecum where the moults occur; the adults inhabit the appendix, caecum and rectum.

Each of the preceding species has at least part of its life cycle outside the host, although the reliance on free-living stages steadily decreases from a complete free-living cycle in *Strongyloides* to the first-stage larvae enclosed in the egg of *Enterobius*. This trend is completed in the case of *Trichinella*

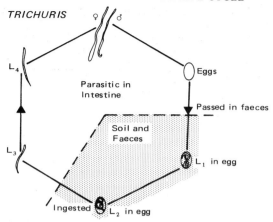

Figure 7.6. Diagrammatic life-cycle of *Trichuris.*

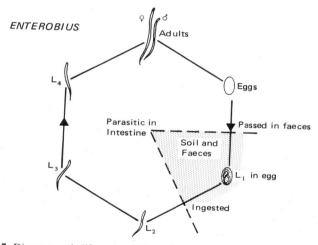

Figure 7.7. Diagrammatic life-cycle of *Enterobius.*

spiralis (figure 7.8). Adult *Trichinella* live in the intestine; the females do not lay eggs, but burrow into the mucosa and release first-stage larvae ovo-viviparously. These larvae penetrate the gut wall and migrate to voluntary muscles where they encyst. They undergo no other development until the flesh is ingested by a carnivore, when the larvae excyst in the intestine and moult to give rise to the adults. Man obtains infections from eating under-cooked pork containing encysted larvae. As man is not regularly preyed upon, he tends to be a "dead-end" host, but *Trichinella* is one of those

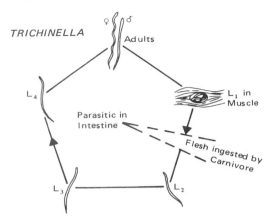

Figure 7.8. Diagrammatic life-cycle of *Trichinella*.

parasites that is catholic in its choice of host; it can develop in such diverse mammals as mice and seals, and is common among all flesh- and carrion-eating mammals. The tendency to reduce the importance of free-living stages is thus completed, and *Trichinella* remains parasitic throughout its life cycle.

The nematode parasites of vertebrates

As stated earlier, the previous species are not related phylogenetically and have been somewhat artificially selected to illustrate some of the trends identifiable amongst nematode life-cycles. The nematodes have left little appreciable fossil record, so that phylogenetic evolutionary patterns are difficult to establish. However, morphological relationships are clearly visible, and Anderson (1957) has shown that it is possible to trace trends in life-cycle patterns amongst a single family of filarial nematodes, the Dipetalonematidae, and thus gain an insight into phylogenetic relationships.

The best-known genera of the Dipetalonematidae are probably *Dirofilaria* and *Onchocerca*. The adults of these nematodes inhabit the heart and subcutaneous tissues of vertebrates respectively. The females do not lay eggs, but release early larvae known as "microfilariae". Microfilariae are released into the cirulatory system or dermal tissue, and do not develop further till ingested by a haematophagous insect. Within the insect the nematode larval stages occur, resulting in an infective third-stage larva that is introduced into a new host when the insect feeds.

It is generally accepted that parasitic nematodes have evolved from free-living ancestors, and it is obviously a massive step from a free-living existence to one involving two hosts and no free-living stage at all. We have already seen, in the first part of this chapter, a series of stages that different nematodes may have passed through in establishing themselves as animal parasites. What Anderson has been able to do is to suggest the steps that may have been taken by representative species of a single family. The stages are illustrated diagrammatically in figure 7.9.

It is postulated that the Spiruroidea and Filaroidea both developed from a common ancestor which inhabited the intestine. Eggs passed in the faeces were ingested by a coprophagous arthropod, where they developed and infected another vertebrate when the arthropod itself was ingested more or less accidentally. Such a life cycle is found in the fish nematode *Rhabdochona ovifilamenta*, which is transmitted by the fresh water amphipod *Hayalella*.

This typically spirurid life-cycle is retained in the bird parasite *Oxyspirura*, but the adults inhabit the orbit, and the eggs pass via the lachrymal and nasal ducts to the gut, and thence into the faeces. The eggs hatch when ingested by a cockroach, and the released first-stage larvae penetrate the gut wall, moult twice in the body cavity, and the third-stage larvae encyst there. When a bird eats the cockroach, the larvae excyst in the crop, and the fourth-stage larvae migrate to the orbit, where they become adult. Two hosts have thus become established in a terrestrial life-cycle, but there is still a short stage spent in the soil. This short free-living stage has been lost by species of *Thelazia* which parasitize the eyes of dogs, sheep, cattle and horses, and very occasionally man. The adults live in the conjunctival sacs. Eggs are released into the lachrymal fluid and hatch to give the first-stage larvae; they are ingested by muscid flies which cluster around the eyes. The larvae penetrate the gut wall and enter the ovaries, where they moult to give third-stage infective larvae. These migrate to the fly's mouth parts and are released into the lachrymal fluid when the fly visits a fresh host. The fourth-stage larvae migrate to the conjunctival sacs and lachrymal ducts, and there develop to adulthood.

Adult *Thelazia* have a habit of migrating in the orbit and the surrounding ducts and tissues. Indeed it is this migration of the worms that causes the pathological symptoms of the disease; and it is a relatively easy step to imagine a nematode living in the subcutaneous tissue around the orbit and migrating to the orbit to lay its eggs. This stage remains hypothetical, as no species has been found that fully demonstrates it. However, *Parafilaria multipapillosa*, the causative organism of "summer bleeding" in horses,

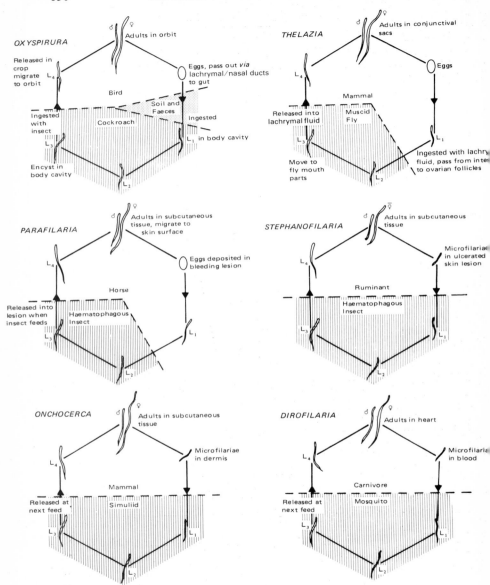

Figure 7.9. Diagrammatic representation of the life cycles of dipetalonematid nematodes, illustrating changes during evolution of the family.

lives subcutaneously and the adult female migrates to the skin surface to lay her eggs. Penetration through the skin causes a bleeding lesion which gives the name to the disease. The eggs are released into the lesion and hatch rapidly to give first-stage larvae. Haematophagous insects are attracted to the site of the lesion; they ingest the larvae which develop in the body cavity and are released into another lesion when the insect again feeds.

The eggs of *Parafilaria* hatch shortly after oviposition, and it is the larvae that are generally ingested by the insect. Egg production does not occur in species of *Stephanofilaria*, and the adult female contains coiled microfilariae within the uteri. The adults live in subcutaneous tissue nodules that enlarge by confluence of several such lesions. The lesions become thickened often break and irritate the host, so that it scratches. The microfilariae are released in the blood which leaves these lesions and are picked up by haematophagous insects attracted to the blood. There is still doubt as to the identity of the intermediate host for many species of *Stephanofilaria*, but muscid flies have been shown to transmit some species. "Hump sore" produced by *Stephanofilaria* sp. is a disease of economic importance in some parts of the U.S.S.R., not only for the distress it causes to the cattle, but also as the lesions damage valuable hides.

In each of the previous species, the nematode stage transmitted from the vertebrate to the arthropod has been released, either into the soil or onto an outside surface of the host to be ingested by the intermediate host. The microfilariae of *Onchocerca* remain in the dermal tissue, where they are deposited by the adult females until ingested by biting flies. The adults live coiled together in sub-dermal nodules (figure 7.10) and the microfilariae migrate through the dermal tissue, e.g. adult *O. gutturosa* in cattle in Britain typically are found within the cervical ligament, whereas the greatest concentration of microfilariae occurs in the skin in the region of the umbilicus, almost as far away as it is possible to get in the same host! The intermediate host of *Onchocerca* is a blackfly of the family Simulidae. These flies have a short scarifying proboscis, ideally suited to breaking through the epidermis and feeding on blood released from ruptured dermal vessels. It appears that the microfilariae may actually migrate towards the site of a *Simulium* bite attracted by a salivary secretion, and thus increase the chance of ingestion by the fly. Development within the fly takes 6–10 days within the thoracic muscles, and the infective larvae migrate forwards to be introduced into the lesion produced when the fly next feeds.

The final stage in this process of life-cycle development is exemplified by *Dirofilaria*. The adults have left the comparatively superficial subcutaneous tissue and live within the heart. Microfilariae are produced, and these

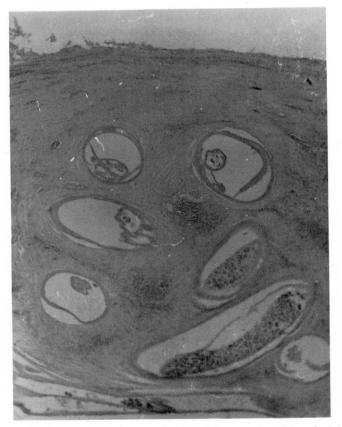

Figure 7.10. Section of an *Onchocerca* nodule showing several sections of adult worms containing microfilariae.

enter the blood circulatory system and pass round the body. Transmission is effected when the microfilariae are ingested by a mosquito. The tubular proboscis of the mosquito is able to probe blood vessels much deeper than the short mouth-parts of *Simulium*, and the microfilariae are sucked in with the blood meal. The infective third-stage larvae are considerably larger than the microfilariae, and distort and probably break out of the proboscis when the mosquito next feeds. Entry into the skin of the new host is thought to be by penetration of the infective larvae through the lesion made by the mosquito rather than by direct injection through the proboscis.

The trend shown by the human parasites discussed earlier, of greater dependence on the parasitic stages and less reliance on a free-living existence, is thus reflected again in the Dipetalonematidae. In these nematodes, however, there is another trend, that of greater specificity of the intermediate host and greater specialization in the mode of transfer.

The great diversity of nematode habit and habitat means that there is a wide range of life cycles, of which the previous dozen or so are but a selected sample, and there are many species whose life cycle does not fall into patterned sequences.

Table 6.1 on p. 104 gives the fecundity of various nematode species. Among the animal parasites, very marked differences occur in the number of eggs produced, although all the species are "successful" parasites. The number of eggs produced reflects the ease with which the life cycle is completed. Those species that rely on eggs as the infective phase, rely on contaminating the food and water of the host to effect transmission and, in general, produce large numbers of long-lived eggs. The strongyles tend to produce less eggs, but behavioural adaptations of the infective larvae allow them to enter the sphere of the host, and thus increase the chance of contact. The case of *Strongyloides* is rather different; the parasitic adults produce relatively few eggs, but auto-infection and the free-living cycle that acts as a multiplicative phase, ensure that infections are maintained.

We have discussed a number of species in which the transmission is via free-living larval stages with the adults parasitic. Some nematodes have the opposite arrangement, in which the adults are free-living and the larval stages parasitic. The best-known examples of this type of life cycle are amongst the insect parasitic mermithids. The adults are large free-living nematodes that live in the soil. When gravid, the females ascend the vegetation and deposit eggs which have attachment threads on them and stick to the leaves. Having laid her eggs, the female dies, and no further development occurs unless the eggs are eaten by a suitable insect such as a grasshopper. The eggs hatch in the insect gut, and the larval nematode passes into the haemocoel; here it moults three times to give a fourth-stage larva that is considerably longer than the host insect (figure 8.4). When fully developed this "post-parasitic" larva leaves its host by penetrating the host integument—a process that normally results in the death of the host—and enters the soil. The free-living stage may last for up to 18 months before the adults are fully mature and ready to lay fresh eggs.

Scientists in general, and biologists in particular, delight in categorizing and classifying. However, living systems do not always oblige by fitting into their inevitably artifical systems of classification. Nematologists have a habit of thinking in terms of free-living and plant or animal-parasitic nematodes. This simplistic view is confounded by the description by Goodey (1930) of the rather remarkable life-cycle of *Tylenchinema oscinellae,* a nematode parasitic in both oat plants and also the frit-fly *Oscinella frit* that attacks oats. The larval nematodes are found within the oat tissues, where they moult to give rise to adult males and females; mating occurs and the inseminated females enter frit-fly larvae. They remain within the insect during pupation and emergence, by which time they have increased in size and are coiled within the body cavity. Large numbers of larvae are produced ovoviviparously, and these fill the body cavity. The larvae moult at least once and probably twice, and then penetrate the gut wall in order to reach the exterior via the anus. The presence of the nematode causes sterilization of the fly. However, this does not affect the fly's behaviour, and it goes through the motions of depositing eggs on a fresh oat plant. At this time the nematode larvae emerge, enter the plant and find their way to fly larvae; here they undergo the final moults to give new adults.

This type of life cycle has rarely been described, but it does illustrate the considerable plasticity of the same basic life-cycle pattern, and with the previous examples emphasizes the great range of possibilities that nematodes have developed. Details of the more bizarre life-cycles are poorly understood, and there is little doubt that many more fascinating examples remain to be described.

CHAPTER EIGHT

THE BIOLOGY OF PARASITIC NEMATODES

A GREAT DEAL OF KNOWLEDGE ABOUT PARASITIC NEMATODES HAS accumulated and any discussion of the biology of nematodes must inevitably include extensive reference to parasitic species. The biology of parasitic nematodes is inextricably linked to that of their hosts. This chapter is an attempt to present, in very truncated form, some of the factors involved in these relationships.

Seasonality and periodicity

Under optimal conditions, free-living microbivorous nematodes such as *Caenorhabditis elegans* or *Panagrellus redivivus* maintain a generation time of a few days or, at most, two or three weeks. This continues as long as conditions permit and, because there are sufficient favourable microhabitats, and the eggs and larvae may be carried by agents such as water, wind, insects and soil, they survive all over the world. If appropriate conditions develop, colonizing nematodes usually appear very soon. Ecologists term such life strategies *r*-strategies: they are characterized by organisms with short generation times and great fecundity, which can rapidly exploit unstable situations. Not all nematodes show this type of strategy. The marine nematode *Enoplus* has an annual cycle, and the plant-parasite *Xiphinema* takes three years to mature.

At first sight, parasitic nematodes may appear to be *r*-strategists, as they are well known to have prodigious reproductive capacities. Many, however, have become modified to synchronize their reproductive effort with the biological habits of their hosts and with their climatic macroenvironment. As knowledge increases, it is becoming apparent that such modifications are widespread amongst nematodes, but the mechanisms which control them are not fully understood.

One of the most certain ways of infecting new hosts is for parasites in female hosts to be transmitted directly to their offspring. Examples of this transmission cycle are common among protozoan parasites, and there are representative cases in the nematodes. Larvae of *Toxocara canis*, the common gut ascarid of dogs, are activated in pregnant bitches to pass into the tissues and placenta, and thus infect the pups *in utero*. Hookworms and other nematodes may use the placental route to augment their more normal cycle of transmission. There are also cases of larval nematodes being carried to suckling young in the milk of lactating females.

The release of eggs or larvae from the host is related not only to their production by the females, but also to the defaecation habits of their hosts. Phillipson (1974) has shown that the passage of pellets by mice reaches a peak about sunrise, and that the passage of the eggs of the oxyurid parasite *Aspiculuris tetraptera* also reaches a peak at this time (figure 8.1).

Adult female filarial nematodes live in the blood, connective tissue and deeper tissues of vertebrates, where they release microfilariae. Microfilar-

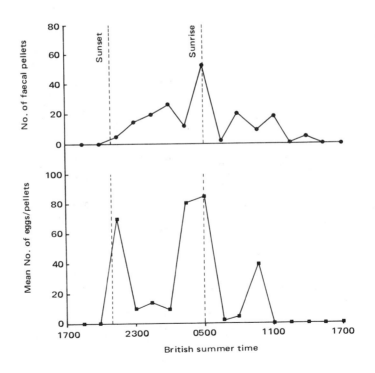

Figure 8.1. Production of faecal pellets and *Aspiculuris tetraptera* eggs in mice (redrawn from Phillipson, 1974).

iae are very small first-stage larvae with a diameter no bigger than an erythrocyte (figure 8.2). These pass to the skin through capillaries or connective tissue, and are taken up in the meal of a blood-sucking arthropod. In the human form which causes elephantiasis, *Wuchereria bancrofti*, the microfilariae move throughout the circulatory system.

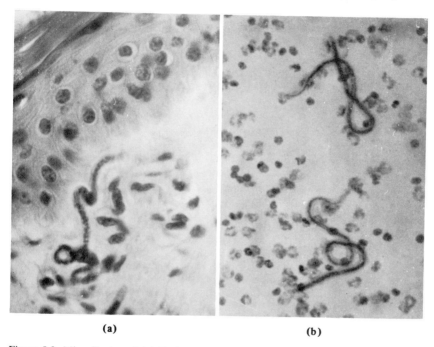

(a)   **(b)**

Figure 8.2. Microfilariae of *(a) Onchocerca volvulus* in skin, *(b) Wuchereria bancrofti* in blood.

During the day they are concentrated in the lungs, owing to some "retention factor" (Hawking, 1967); but at night they are more evenly distributed, and consequently the number in the peripheral blood rises. The vectors are night-biting mosquitoes, such as *Culex fatigans,* so the circadian periodicity of the microfilariae coincides with the biting habits of the vector. The African eyeworm *Loa loa* shows a reversed periodicity. This human parasite is transmitted by the day-feeding blood-sucking fly *Chrysops,* and the microfilariae are found in the greatest numbers in the peripheral blood at midday. The simian strain of the same parasite shows nocturnal periodicity, and the vectors are night-biting species of *Chrysops.*

Pathology of nematode infections

The concept of disease

"Disease" is a very complex phenomenon. Parasitic nematodes can cause pathology or a reduction in the biological efficacy of an infected individual, compared with non-infected members of the same species. There is an element of truth in the "old chestnuts":

Parasites do not kill the goose that lays the golden egg.
Parasites live as blackmailers and not as robbers.

Nevertheless these sayings are too simple to be used as a rule. For ex-

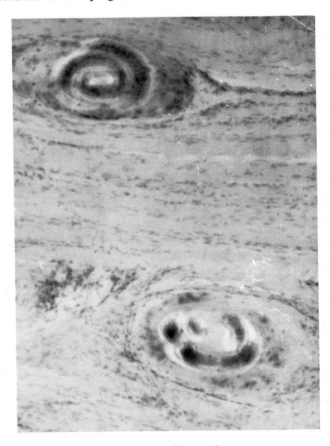

Figure 8.3. *Trichinella spiralis* larvae encysted in muscle.

ample, in the life cycle of *Dracunculus medinensis* the infection is carried to man when he ingests infected copepods with drinking water. The whole biology of larval *D. medinensis* is directed towards the death of the copepod. Infective larvae of *Trichinella spiralis* enter the muscles of their host, and remain until the host flesh is devoured (figure 8.3). It seems good biology to interpret the "disease" of these infected hosts as enhancing the possibility of transmission through the death of the host.

The justification for the above statements is deduced from the fact that, *if* a parasite species was so pathogenic that it destroyed its host populations, then the parasite itself would be destroyed. All existing nematode parasites have clearly evolved to reach some kind of equilibrium with their hosts, in which neither host nor parasite disappear.

Disease can be the direct result of losing nutrients to parasites, but it is just as frequently due to cytological changes induced by the parasites, or to the production of toxins, antienzymes and other antigenic products. Many parasites move on or in their hosts, frequently changing their feeding sites. This can lead to secondary infections from invasion by bacteria, viruses or fungi. There are also examples of nematodes that transmit other parasites from host to host. *Heterakis gallinarum* of fowl carries the protozoan *Histomonas meleagridis*, the causative agent of "Black head" disease; *Aphelenchoides ritzemabosi* on strawberries transmits the bacterium *Corynebacterium fascians*, which causes the condition known as "cauliflower head"; species of the plant parasites *Xiphinema* and *Longidorus* transmit viruses to their hosts. In all these cases the diseases caused by the secondary pathogens can be more serious than those caused by the nematodes themselves.

Many of the so-called "ectoparasitic" plant nematodes are better considered to be "microbrowsers" or even herbivores than true parasites. Biologically they are very similar to aphids or phytophagous bugs that take the juices of plants. The length of time spent at a single feeding site is comparatively short; the nematodes do not enter the plant tissue, but ingest cell fluids through the hollow lumen of the inserted stylet. Once feeding is completed, the stylet is withdrawn and the nematodes leave the site. The damage which can be directly attributed to a single nematode is very slight, but the effect of a large population feeding on a single plant is, of course, cumulative.

Larvae of the cyst nematodes (*Heterodera* spp.) and root knot nematodes (*Meloidogyne* spp.) invade the deeper tissues of their hosts. Entry is effected by the larvae cutting through the cell walls and migrating intracellularly until they settle at a permanent feeding site. The nematodes can induce the plant to form "giant cells' or "syncytia" (figure 5.7) on which they feed. Jones and Northcote (1972) have presented light- and electron-microscope evidence that the syncytium acts as a transfer cell, channelling nutrients from the vascular system of the plant to satisfy the nutritional requirements of the parasite. If the induced syncytium is sufficiently large, the individual nematode becomes a sedentary female; if not, it develops into a male (Trudgill, 1967). Females enlarge greatly and, in doing so, they break through the root wall. The anterior region, however, remains embedded in the giant cell.

Not all endoparasitic plant nematodes remain static and in association with modified host cells. *Pratylenchus* spp., for example, are endoparasites with a very wide range of hosts, including tobacco, potato and daffodil. The adults move through the cortical cells of the roots, feeding on the cell sap and depositing eggs as they progress through the tissues; the migrations are marked by a trail of destroyed cortical cells and lesions in the cortex. These lesions lead directly to the destruction of small rootlets, and leave a series of weakened areas and a generally weakened plant which is very vulnerable to secondary invasion by bacteria or fungi.

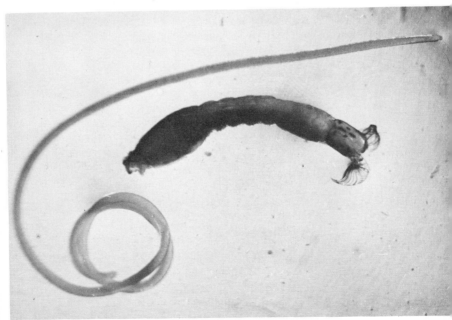

Figure 8.4. Fourth-stage larva of *Hydromermis* and its *Simulium* larva host.

Some parasites are almost invariably fatal to their hosts. The mermithid nematodes of insects spend much of their larval life within the haemocoel, and attain a mass which is enormous relative to that of their host (figure 8.4). The fourth-stage larvae emerge from the insect and, during emergence, tear a hole through the insect body wall which leads to the death of the host. The fact that the host dies has led to mermithid nematodes being used as biological control agents. The situation in which a single nematode individual causes the death of its host is unusual; most species have a much more equitable relationship with their hosts.

Site specificity

Host tissues present a series of widely contrasted microenvironments and, in common with other parasites, nematodes are restricted to specific sites.

Table 8.1. Nematode parasites recovered from the alimentary tract of a single lamb at autopsy.

Region of gut	No. of nematodes	Nematodes
Stomach, Abomasum	8,500	Haemonchus
Small intestine		Ostertagia
Duodenum	700	Ostertagia Trichostrongylus
Jejunum	11,000	Ostertagia Trichostongylus Bunostomum Cooperia
Upper ileum	1,850	Cooperia Nematodirus
Mid ileum	2,050	Nematodirus Strongyloides
Lower ileum	90	Strongyloides Trichuris
Caecum	750	Trichuris Oesophagostomum
Large intestine	750	Chabertia
Total	25,690	

Dictyocaulus viviparus lives as an adult in the lungs of cattle; *Dioctophyme renale* inhabits the kidneys of dogs; *Capillaria hepatica* lives in rodent livers, and so on; every species of nematode is characteristically located in a specific site.

Adult gastro-intestinal parasites often occupy a single localized site within the gut. Table 8.1 lists the nematodes found at post-mortem of a single lamb. Some species overlap with others, but it can be seen that there is a general progression through the tract. The table also illustrates that related species tend to occupy adjacent situations, and that normal infections can involve thousands of individuals.

Pathology

Adult *Ascaris lumbricoides* are lumen dwellers. In the empty gut they probably occupy a great deal of the cross-sectional area but, when it is full, they are normally found near the mucosa. *A. lumbricoides* feeds on

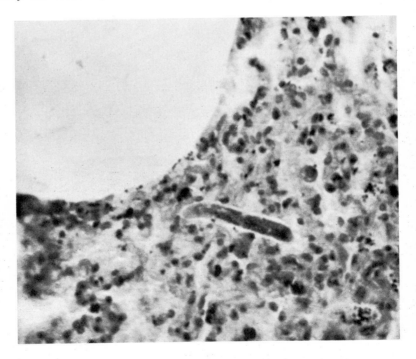

Figure 8.5. Migrating *Ascaris lumbricoides* larva in the lung.

undigested or partially digested host gut contents. Concentrations of mucosal epithelial cells have been seen in its gut, so it may take these preferentially. The pathology of *A. lumbricoides* infections is normally related to the mechanical and physical effects of large volumes of worms (up to 10 lb of *Ascaris* in a single host!). A heavy infection can cause blockage of the gut. Adult *A. lumbricoides* are notorious for a propensity to wander from their site in the gut, and may emerge from the anus, nostril

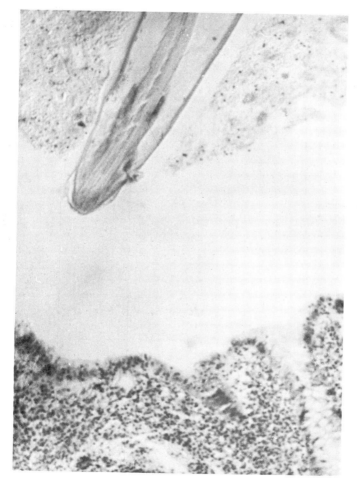

Figure 8.6. Anterior of *Enterobius vermicularis* within the lumen of the appendix.

or mouth, when they probably cause more shock and distress than disease. Such movements, however, can be much more serious and, although uncommon, *A. lumbricoides* has been seen to emerge from the orbit and from the ear, after migrating into the lachrymal duct and auditory tube respectively. More common is the migration of *Ascaris* into the gall bladder and biliary ducts, resulting in jaundice, or into the appendix, where it may initiate appendicitis.

Ascaris illustrates another feature typical of many nematode infections of man and animals, namely that the larval stages of the parasite may be more significant in causing disease than the adults. After ingestion, *Ascaris* eggs hatch in the gut, and the larvae penetrate the gut wall and move to the

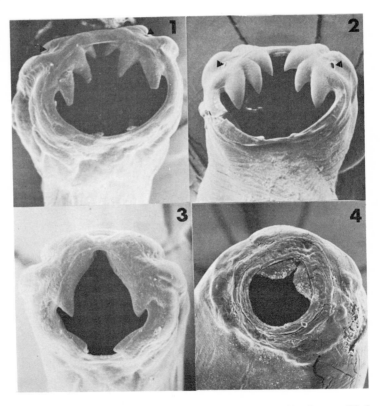

Figure 8.7. Stereoscan photographs of the stoma of four species of hookworm. (1) *Ancylostoma caninum* from dog, (2) *A. tubaeforme* from cat, (3) *A. ceylanicum* from cat, (4) *Necator americanus* from man (1–3 Matthews and Croll, 1974).

liver. Following a developmental phase in the liver, they migrate to the lungs and burst through the alveoli into the lumen (figure 8.5). If large numbers of eggs are ingested on a single occasion, the migration of the larvae through the lungs is synchronized and may cause severe "ascaris pneumonia". Serious human disease resulting from *Trichinella spiralis*, *Toxocara canis* and *Anisakis* are all due to the larval stages undergoing tissue migrations.

Enterobius vermicularis is another lumen-dwelling intestinal parasite that has been incriminated in the etiology of appendicitis (figure 8.6). In this instance the evidence is less clear-cut, as the nematodes are normally found in the lumen and do not generally invade the mucosa; also, *E. vermicularis* is found in normal appendices as well as those showing chronic or acute inflammation. Beyond the possible involvement in appendicitis, *Enterobius* is a relatively innocuous parasite that rarely, if ever, produces serious lesions. The same is not true of the hookworms (figure 8.7) and the blood-feeding strongyles and trichostrongyles which remain closely attached to the mucosa.

Hookworms may cause two distinct types of disease. In susceptible hosts, skin penetration is followed by entry into the blood or lymphatic systems and a migration, similar to that of *Ascaris*, to the intestine. The migration of larval hookworms through the lungs may result in local lesions but, unless the infection is exceptionally heavy, these are rarely

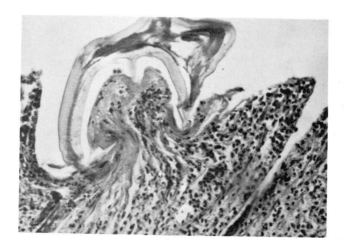

Figure 8.8. Adult *Ancylostoma tubaeforme* attached to the gut mucosa.

more than transient, and the main pathology of hookworm infections is due to the adults in the intestine.

Adult hookworms pull a plug of the gut mucosa into their buccal capsule (figure 8.8) and suck blood from the macerated mucosa. The nematodes may move from one site to another; and each time they change position, they leave haemorrhagic areas that are liable to secondary bacterial invasion. In man, if the host is well nourished, infections with 50 or less hookworms normally have little effect but, as the infection level increases, the blood haemoglobin falls, and severe anaemia results from infections of 450 or more adults. The degree of anaemia from a given infection is dependent on a combination of dietary iron intake and hookworm blood loss. Hookworm disease is thus much more serious in those patients on an inadequate diet.

Hookworm larvae that enter hosts in which further development cannot, or does not, occur, may remain in the skin and give rise to the condition known as "creeping eruption" or "cutaneous larval migrans". Most cases of creeping eruption in man have been attributed to *Ancylostoma braziliensis* of canine origin, and it has generally been assumed that the condition is caused by non-human hookworms entering man. Maplestone (1933) and Beaver (1945) both reported typical creeping lesions following multiple exposure to invasion by the human hookworms *A. duodenale* and *Necator americanus*. They suggested that the lesions were largely a response following sensitization. Without doubt secondary bacterial invasion often contributes to the severity of the lesions, but mostly they are manifestations of the host immune response to the migrating larvae.

Human infections with animal ascarids, especially *Toxocara canis,* have been termed "visceral larval migrans". These infections are especially common in children, when the association with dogs and the ingestion of soil contaminated with infective eggs, play an important role in the epidemiology. The eggs hatch in the intestine and migrate into many organs, especially the liver. Maturation of the larvae does not occur in man and the larvae continue to migrate around the body. The lesions that follow the larval migrations are most serious when they occur in the nervous system, and especially the retina where they may cause blindness.

Gastro-intestinal nematodes cause great losses among domestic animals. The main clinical features associated with these infections are diarrhoea, loss of appetite with consequent loss of weight, and reduction in food utilization. Lambs infected with *Trichostrongylus colubriformis,* for example, require twice the weight of feed to produce 100 lb live weight than is required by controls; this unthriftiness may also lead to a later reduction in wool production. Heavy infections may lead to death, especially in young animals. Each species of parasite shows a characteristic pattern of pathological processes. These have been best studied in *Trichostrongylus* spp. and *Haemonchus contortus* in sheep, and *Ostertagia* spp. in sheep and cattle.

Infective larvae of *Ostertagia* are ingested during grazing, and the larvae that enter the stomach penetrate the high mucus-secreting glands. They remain in the glands, moult and mature. At the end of the third week after infection, they emerge to become adults on the stomach mucosa. The invaded glands become distended, and after 16–18 days the cells lose their morphological and functional differentiation. In heavy infections the lesions formed by individual parasites merge to form an intensely hyperplastic mucosa that may lose all its normal function. Emergence of the larvae from the glands is followed by death of the superficial cells and sloughing of the epithelium. The adult worms lie on top of the damaged mucosa, embedded in an exudate of protein and cells.

Pepsin is essential for the efficient functioning of the digestive processes in the stomach. Pepsinogen, produced deep in the glands, is normally activated by HCl in the lumen to form pepsin which only acts at a low pH. The loss of effective functioning of the gland cells results in a reduction in the production of HCl, and the abomasal pH rises (figure 8.9). This is at the time in the infection that the larvae are emerging from the glands, and is also the time when severe diarrhoea occurs. It is thus apparent that the clinical features can be directly related to the pathological processes.

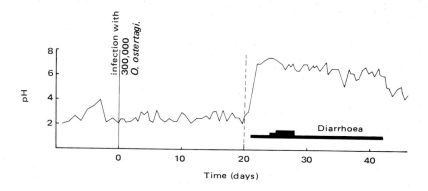

Figure 8.9. Changes in the pH of the gastric juice during a single cycle of infection with *Ostertagia ostertagi* in a cannulated calf (redrawn after Jarrett, 1966).

The lung-worms, *Dictyocaulus* spp. of cattle and sheep and *Metastrongylus* of pigs, mature in the lungs of their respective hosts. *Dictyocaulus* has a direct life-cycle and the infective larvae are ingested with grass. The infective stage of *Metastrongylus* is reached in an annelid intermediate host and reaches the pig's intestine when the earthworm is eaten. The larvae penetrate the host gut and pass via the blood stream to the lungs, where they mature. The effects of lung-worms in pigs are not normally serious, although the cough, bronchitis and secondary pneumonia that they cause, may prove fatal to young animals. In cattle and sheep, the effects of *Dictyocaulus* while similar in principle tend to be more severe in practice. During the first week of the infection, larvae penetrate the gut and migrate to the lungs; this phase does not normally give rise to clinical symptoms. It is during the prepatent maturation phase within the lungs that the most serious effects of the parasite are manifest. The larvae cause blockage of the bronchi and bronchioles, and this may result in the collapse of alveoli. A great deal of mucus is produced, and this becomes mixed with air and blood from the pulmonary lesions. Acute pneumonia is a frequent complication, and bacterial invasion commonly occurs. The critical phase of the disease lasts for 6–7 weeks, after which there is a gradual recovery. Hosts that recover from an infection with *Dictyocaulus* retain a long-lasting immunity to further reinfection. This fact has led to the development of an efficient vaccine using X-irradiated attenuated *Dictyocaulus viviparus* larvae to stimulate a host response.

Resistance

The host is not without the means of combating the effects of parasite attack. Some species, strains or even individual hosts may be resistant to a given nematode species, and some strains of parasite may be more infective than others. Although some of these variations are obviously due to physical factors, the biological reasons underlying the differences are poorly understood. *Rotylenchulus reniformis* has been found to attack more than 200 species of plant, many of which are of economic importance (Ayala and Ramirez, 1964). In many of these, although large numbers of nematodes attack the roots, there is no growth reduction. However, in other plants, e.g. cotton, considerable damage is done and the growth is much reduced.

One of the more remarkable mechanisms of resistance is the so-called "self-cure" reaction. Although first described by Stoll (1929a,b) for *Haemonchus contortus* in lambs, the self-cure phenomenon has since been shown to occur in other species, including *Trichostrongylus* spp. in rabbits and *Nippostrongylus brasiliensis* in rats. As typified by *H. contortus*, "self-cure" consists of the rejection of a patent infection by a subsequent invasion of larvae of the same or a closely related species. Following "self-cure" the host is apparently refractory to further reinfection.

Haemonchus contortus lives in the abomasum of ruminants, and feeds on blood ingested from the host mucosa. Infective third-stage larvae are ingested and burrow into the mucosa, where they moult to the fourth stage. These remain in the mucosa in contact with the host blood vessels before moulting to the adult stage and passing back to the gut lumen. During their time within the mucosa, the fourth-stage larvae cause a marked increase in the histamine level in the blood, and also induce a rapid synthesis of antibodies. Following a primary infection, the histamine and antibody production are not sufficient to affect the subsequent blood feeding of the adults. However, a challenge infection results in a marked increase in the circulating antibodies, which results in the established adults being removed from the gut wall and appearing in the faeces. Under pasture conditions there is almost continuous exposure to infection, and consequently a steady supply of fourth-stage larvae within the mucosa. This maintains the circulating antibodies; it prevents the establishment of new adults in the lumen and causes the apparent resistance to re-infection.

Acquired immunity

As so many parasites show host specificity, it has been a widely held view that much of the immunity shown to helminths is "innate". However, recent more sophisticated techniques have illustrated that helminths do induce antibody production by their hosts. Most nematodes have a relatively long life within the host, and thus appear to be able to evade the host immune system. The production of truly protective antibody that results in worm expulsion appears to be rare and, although a number of nematodes induce the formation of antibody that can be used for immuno-diagnosis, this does not always appear to affect the course of a primary infection.

Possibly the best studied nematode host immune system is that of *Nippostrongylus brasiliensis* in the rat. Infective larvae penetrate the skin and, after a migration through the lungs, trachea and stomach, establish themselves in the small intestine. Adults produce eggs on the sixth day, and continue to do so for about a week; after that time, most of the worms are expelled over the next two or three days. If, instead of a single large initial dose of larvae, the rat is exposed to only a small number at daily intervals, a relatively stable adult population is established which may persist for months. The adults in this population are smaller, produce fewer eggs per female, and appear to be physiologically adapted to living in the immune host.

A number of possible theories have been put forward to explain how helminths may be able to avoid the host's production of protective antibody (Ogilvie, 1974).

1. The parasite may produce antigenic substances that are so like host substances that they are not recognized as "foreign", and are thus not immunogenic.
2. The host may produce antibody against antigen that is not important to the parasite, which is thus not affected by the immune reaction.
3. The parasites produce molecules that resemble those of the host and effectively disguise the important antigens from recognition.
4. The helminth responds to immunity by changing the immunogenicity of its antigens.

Experiments transferring *Nippostrongylus brasiliensis* adults, produced from either a single large (normal), or many small (adapted), exposures to infective larvae, to pre-immunized rats, resulted in the rapid expulsion of the normal worms, but a relatively stable population of adapted ones (figure 8.10). A mixed infection of normal and adapted adults transplanted

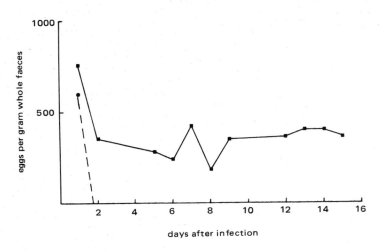

Figure 8.10. Faecal egg counts taken at intervals from pre-immunized rats with 200 "normal" ●————● and 200 "adapted" ■————■ *Nippostrongylus brasiliensis* worms after laparotomy (redrawn from Jenkins and Phillipson, 1972).

Table 8.2. *Nippostrongylus brasiliensis* recovered from previously uninfected rats with a mixed population of 50 "normal" and 50 "adapted" male and 100 "normal" and 100 "adapted" females.

| Day killed after infection | Worms recovered (± mean) | | |
	Normal and adapted males	Normal females	Adapted females
12	83 ± 15	60 ± 16	71 ± 20
42	75 ± 11	27 ± 14	71 ± 17

into clean rats resulted in the rejection of the normal nematodes after the usual 14 days, but the infection persisted with the adapted adults (Table 8.2) (Jenkins and Phillipson, 1972).

It is possible to transfer some immunity to *N. brasiliensis* to clean rats passively by either immune serum or immune lymphoid cells, but the response is less than that found in actively immunized animals. Love (1975) has demonstrated that the proper functioning of the immune system requires the interaction of both immune serum and sensitized lymphoid cells. When both were injected into normal rats, transplanted *N. brasiliensis* were rejected at a rate equivalent to that seen in actively immunized rats. Fourth-stage larvae were more susceptible than adults to the effects of passive immunity, but it is suggested that the mechanism for expulsion of adult worms, following an active primary infection, and larval worms in a pre-immunised host, is similar. Antibodies in the serum probably damage the nematodes, which are then susceptible to the action of the sensitized lymphocytes.

The host response to an infection of *N. brasiliensis* is relatively rapid, and expulsion of normal large invasions is effected within three weeks, after which the rats have a long-lasting immunity to subsequent infections. Similar responses have been shown for other nematodes, but the time scale may be more protracted.

CHAPTER NINE

SURVIVAL OF NEMATODES

ALL ORGANISMS HAVE A FINITE LIFE SPAN. EACH PARASITE IS THUS faced with the problem of transmission from one host to another if the species is to be perpetuated. To achieve this transmission, special stages are produced. For most nematodes the transmission stage is an egg, although some may ovoviviparously produce first-stage larvae, as in *Trichinella,* or microfilariae, as in *Wuchereria* and *Onchocerca.* Infective stages have a period outside their definitive host, either free-living or in an intermediate host. Free-living stages must withstand the vagaries of the environment, and we have already seen how the structure of the egg shell may

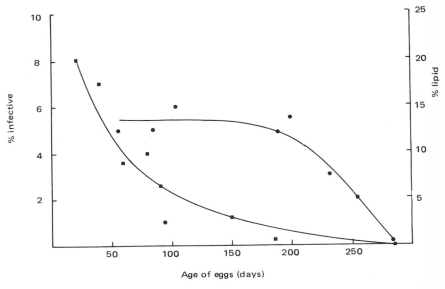

Figure 9.1. Decrease in infectivity ●————● and lipid level ■————■ of *Ascaridia galli* eggs with time (redrawn from Elliott, 1953).

affect the ability of the developing larva to resist environmental changes. Although the infective stage may remain viable, infectivity decreases with time.

The infective eggs of *Ascaris lumbricoides* may remain viable and infective for at least six years. Unwashed strawberries grown on ground experimentally contaminated with about 3000 *A. lumbricoides* eggs per square yard were eaten each year for six years by two volunteers, and each year a light *Ascaris* infection was contracted. The exact numbers of eggs ingested were not recorded however, and quantitative analyses were thus not possible.

Elliott (1953) fed known numbers of infective eggs of *Ascaridia galli*, aged for different lengths of time, to two-week-old chicks. The infectivity of the eggs was assessed from the number that had developed to adults and were found at autopsy three weeks later. The percentage recovery was low

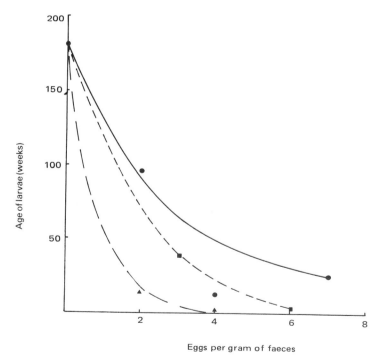

Figure 9.2. Mean daily egg counts from goats, four weeks after infection with *Haemonchus contortus* third-stage larvae aged at 7°C ●————●, 24°C ■--------■, and 37°C ▲————▲ (redrawn from Rogers, 1940).

at all ages, but the picture that emerged was that the eggs retained their infectivity for a period of about 200 days, after which it fell rapidly away (figure 9.1). The main food reserve of the larvae is lipid, and the lipid level of *A. galli* artificially hatched from their eggs fell exponentially with time (figure 9.1).

The majority of studies on nematode infectivity have been conducted using those nematodes whose eggs hatch to give free-living larvae. Known numbers of aged eggs and larvae have been fed to susceptible worm-free hosts. The number of eggs released per gram of faeces produced, or the number of adults recovered at autopsy, after an appropriate time for maturation, has been used to assess infectivity. Rogers (1940) and Rose (1963) have used this method to investigate the effect of age on infectivity of third-stage larvae of *Haemonchus contortus* in goats and sheep respectively. Rogers (1940) found a rapid decrease in infectivity during the first 2–3 weeks, after which the rate of decline slowed (figure 9.2). The longevity of the larvae as assessed by their infectivity was markedly influenced by the temperature at which they were stored. The sheep is perhaps a more natural host for *H. contortus* than the goat, and this may explain why Rose (1963) found larvae still infective after 8½ months. These larvae had been aged under simulated field conditions rather than at a constant temperature. Larvae stored at constant temperature showed a maximum longevity that was inversely proportional to the storage temperature (Table 9.1).

The role of activity in the penetration of hookworm larvae has recently been emphasized (Matthews, 1972, 1975), and nematodes that invade their hosts by penetration through the integument rely upon activity to effect entry. As only those larvae that are able to penetrate the skin have the opportunity of maturing, *in vitro* measures of penetrability may be used to estimate the infectivity of larvae. These methods have all been modifica-

Table 9.1. Maximum longevity of *Haemonchus contortus* larvae, aged at different temperatures (Rose, 1963)

Temperature (°C)	Maximum Longevity (weeks)
24–25	35
15–16	49
10–11	83
4–5	87

tions of Goodey's floating-raft technique. A suitable skin membrane is suspended with the dermal surface in intimate contact with isotonic saline at body temperature. Infective larvae introduced on to the top skin surface are allowed to penetrate for an appropriate time, and those that do so are collected from the saline.

Quantitative studies using this technique to investigate the relationship between ageing and infectivity have been conducted on infective larvae of *Strongyloides ratti* (Barrett, 1969) and species of cat hookworm (Rogers, 1939; Croll and Matthews, 1973). As larvae aged, the percentage penetrating fell (figure 9.3). The activity of the larvae, measured as the number of larval undulations per minute, fell in a similar manner. As only motile larvae are able to penetrate, the rate of movement above a critical threshold is probably of secondary importance in effecting entry.

As larvae age, their lipid food reserve decreases, and many authors have suggested that the "physiological" rather than the chronological age of the larvae can be measured by the level of lipid. The loss of lipid has been related to the decrease in infectivity. Barrett (1969) found a decrease in respiratory rate $Q(O_2)$ of infective *S. ratti* larvae with age, although the respiratory quotient (R.Q.) remained the same. The low R.Q. suggested that lipid rather than glycogen was being metabolized, and that there was no change in the substrate utilized with age. Costello and Grollman (1958) collected the results of a number of workers relating the $Q(O_2)$ and

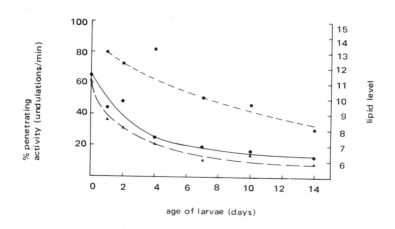

age of larvae (days)

Figure 9.3. Effect of ageing on activity ▲—— ——▲, penetration ●————● and lipid level ■- - - - -■ of *Ancylostoma tubaeforme* infective larvae (from Croll and Matthews, 1973).

Table 9.2. Relationship of $Q(O_2)$ to longevity in nematode larvae (Costello and Grollman, 1958).

Nematode	Type of larva	Longevity	$Q(O_2)$
Strongyloides papillosus	Infective filariform	7 days	28.5
Nippostrongylus brasiliensis		30 days	18.4
Haemonchus contortus		90 days	12.6
Trichinella spiralis	Encysted	5 years	2.35
Eustrongylides ignotus		4 years	0.56

longevity of nematode larvae, and an inverse relationship appeared (Table 9.2).

An analysis of the activity and lipid level of hookworm larvae aged under different conditions suggested that ageing was independent of the residual lipid (Croll and Matthews, 1973). Larvae stored at 10°C lost their lipid more slowly than those at 26°C, but their activity at a given age was no greater. The conclusion reached was that the age-dependent phenomena were due to a decrease in metabolic rate. This conclusion receives support from the investigations of Gershon and co-workers in Israel (Gershon and Gershon, 1970). Working with axenically raised *Turbatrix aceti,* they found a reduction in the activity of acetyl cholinesterase, α-amylase and malic dehydrogenase with age. Using immunochemical techniques it was shown that, as the nematodes aged, there was a decrease in the functional protein and an increase in non-functional protein. Van Gundy *et al.* (1967) found a reduction in the esterase and acid phosphatase enzymes that hydrolyse lipids in *Meloidogyne javanica.* There is thus a growing body of evidence that suggests that ageing in nematodes is a manifestation of a reduction in activity of certain critical enzyme systems.

The longevity of nematodes varies widely, and a number of species appear able to survive for extended periods by undergoing some form of reduced or modified metabolism. In none of the reports cited above were skin-penetrating infective larvae found to be active and infective for more than ten weeks; however, Norris (1971) recovered living *Ancylostoma tubaeforme* infective larvae from the tissues of mice 3, 5 and 10 months after infection. When fed to fresh mice, the 5-month-old larvae were found to have retained their infectivity. *A. braziliensis* larvae, digested from mice infected 9 months earlier and fed to worm-free kittens, produced patent infections. Larvae within the mice did not "age" in the physiological sense to the same extent as those kept *in vitro*. Larvae of *Uncinaria lucasi*, a hookworm of the fur seal *Callorhinus ursinus*, remain viable and infective within the tissues of their definitive host for at least a year (Olsen and Lyons, 1965). Unlike other hookworms, per-cutaneous infection does not result in adults being produced directly. The larvae migrate to the blubber, and in the females to the mammary glands, but not to the intestine. Infection is by passage from the females to the pups in the milk. These larvae do develop to adulthood and produce eggs that hatch to give infective larvae that, having penetrated the skin of the flippers or been ingested, migrate to the blubber. The larvae that enter males never develop further, and adult hookworms are not found in any seals other than pups.

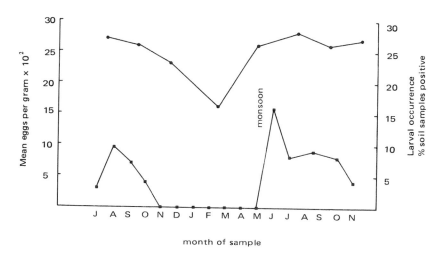

month of sample

Figure 9.4. Relationship between the occurrence of hookworm larvae in soil ■————■ and the intensity of hookworm infection ●————● (from Schad *et al.*, 1973).

Delayed development of hookworm larvae has been reported in man infected with *A. duodenale* in West Bengal, India (Schad *et al.*, 1973). A significant increase in the number of eggs per gram of faeces was found preceding the monsoon. During the previous six to eight weeks, the time required for normal maturation of invading larvae, the occurrence of larvae in the soil was extremely low, and there was no rise that would account for an increased egg output (figure 9.4). The conclusion reached was that the development of larvae acquired during the rainy season of one year is arrested until just before the following monsoon, when maturation is completed.

A seasonal increase in faecal egg counts in sheep has long been associated with lambing, and the early part of the year in the northern hemisphere. The phenomenon has thus been variously termed "spring-rise" or post-parturient rise, and results from *Haemonchus contortus* females producing more eggs after a relative reduction during the winter. Initially it was considered that the rise was due either to a greater fecundity of *H. contortus* females that had overwintered in the host, or that the host immune response was reduced following a smaller worm burden and the stress of winter. Any ingested larvae, although less common on the pasture, were then more likely to mature. The demonstration of the regular association between spring rise and lambing, and the greater extent of the rise in nursing ewes than in virgin ewes and rams, lead to the idea that pregnancy and the hormonal levels related thereto might have contributed to the stress factors responsible for affecting the immune response. Recent studies, however, have illustrated that the parasite is as much responsible as the host for the "spring rise" in egg counts.

During the winter months, infected sheep contain few adult *H. contortus*, whereas large numbers of early fourth-stage larvae are frequently encountered. Blitz and Gibbs (1971a, b) have considered that it is the maturation of these inhibited larvae that constitutes the major supply of adult worms responsible for the spring rise in egg counts. When eggs cultured in the laboratory at different times of the year were fed to lambs, it was found that the percentage of inhibited larvae increased in the August and September infections, but the overall inhibition never reached more than 45%. If, however, the larvae were cultured under constant conditions to the third stage, and then exposed to natural climatic conditions prevailing in September (in Canada where the experiments were performed), 96% of the larvae were inhibited at the fourth stage in a subsequent infection. This compared with 100% inhibition in naturally-grazed lambs. The conclusion was that arrested development was governed by the environmental experience of the infective larvae. This idea receives support from work on the rabbit trichostrongyle *Obeliscoides cuniculi* (Fernando, Stockdal and Ashton, 1971). The percentage of inhibited fourth-stage larvae increased with the length of time that infective third-stage larvae were held at 4°C up to a maximum, after which additional storage decreased the proportion of arrested larvae. At higher temperatures the effect was less marked. Armour and Bruce (1974) have shown very similar results for the arrested development of *Ostertagia ostertagi* in calves. Once again inhibition followed storage of infective larvae at 4°C; a peak was reached after eight weeks storage, then the proportion of inhibited larvae decreased, till by 33 weeks virtually no inhibited larvae were recovered. Freshly harvested larvae used as controls (irrespective of the time of year at which they were isolated) showed no evidence of arrested development. A release of arrested larvae was found between 17 and 18 weeks in two batches of calves infected at different times of the year, suggesting that intrinsic rather than extrinsic factors were responsible for the termination of the period of arrest.

It is thus apparent that some nematodes are able to modify the normal pattern of development to suit the biology of their hosts and their geographical location. *H. contortus* larvae do not survive well during winter on pasture, thus to ensure a ready supply of larvae for the new season's lambs, the state of arrested development has evolved to prolong the larval stages. Arrested larvae mature in the spring, probably as a spontaneous result of the completion of some physiological process. In non-lactating ewes it is suggested that a "self-cure" reaction occurs, so that most of the adults are expelled and the spring rise is relatively small, whereas lactation suppresses the normal immune response and results in the typical "spring rise" phenomenon.

Although the infective larvae of some trichostrongyles may be induced by environmental conditions, such as a reduced temperature, to enter a state of arrested development following a subsequent infection, temperature may play other roles in the control of nematode development. *Nematodirus battus* has a single generation per year, and most adults do not survive the winter within their ruminant host. Infective eggs contain third-stage larvae and do not hatch immediately, but require a period of exposure to reduced temperatures before hatching will occur when the temperature again rises. In temperate regions sufficient cold periods occur only during the winter. All eggs laid during the previous summer release their larvae onto the pasture during a relatively short period when the temperature again rises after the winter. This ensures a large number of infective larvae when the new lambs are most susceptible. Other species of *Nematodirus* have a more typical trichostrongyle life-cycle, and arrested development occurs in some species to overcome the winter conditions.

Arrested development and delayed maturation have now been demonstrated in a wide range of animal parasites. Adult dogs and cats are less commonly found infected with mature *Toxocara canis* and *T. cati* in Britain, but pups and kittens are very commonly infected. A number of age-dependent immune responses may partly explain this, but the major factor is the infection route. When mature eggs are ingested by an adult carnivore, they hatch in the stomach and undergo a typical ascarid migration; however, most larvae do not return to the intestine, and those that do are normally expelled before they become fully developed. The majority of the larvae enter the tissues, and do not develop further unless their host is a pregnant female. In this case the larvae migrate to the umbilical vessels and enter the foetus; thus the neo-natal animal is already infected. The parasites in these young animals mature fully and release eggs. There is some evidence that larvae are stimulated by host hormonal levels to migrate from the tissues to the placenta, but their importance has not been elucidated. *T. canis* eggs may hatch in the intestine of rodents or man. In the former case the larvae migrate into the tissues and remain infective for at least two years—a carnivore eating such an infected "paratenic" host is liable to become infected. In man, ingestion of *T. canis* eggs may give rise to the condition known as "visceral larval migrans". Children are most commonly infected, probably because of their greater tendency to ingest contaminated soil. Migrations of the larvae are normally followed by eosinophilia, but other symptoms are very variable, depending on the level of infection and the organs invaded. Most frequently larvae are found in the liver, lungs, eyes or brain, and it is probably the lesions caused by larvae migrating in the central nervous system to the brain and to eyes that cause the greatest distress. There have been cases of *Toxocara* lesions being confused with retinoblastoma, resulting in the unnecessary removal of the infected eye. Most larvae eventually stop migrating and become encapsulated in the tissue of the visceral organs.

Michel (1974) has recently fully reviewed the literature on arrested development of animal-parasitic nematodes.

Continuous reproduction throughout the year is not possible for the great majority of free-living and plant-parasitic nematodes, as is the case for most invertebrates. Climatic conditions in all but the truly tropical regions preclude the possibility. Cold and dry seasons may intervene between periods of optimal conditions, and most host plants show seasonal abundance. Mammals are relatively longer-lived than the majority of plants and provide more constant conditions for their parasites, but the free-living transmission stages face similar problems of temperature and moisture stress. It is thus not surprising that many species have evolved mechanisms to overcome hazardous environmental conditions.

We have already seen that *Nematodirus battus* is able to pass the winter as a third-stage larva within the egg, and indeed requires exposure to reduced temperature for optimum hatching. It is impossible to make generalizations about the stage in the life cycle or the degree of resistance shown by nematodes to desiccation, or other unfavourable conditions. Fourth-stage larvae of *Ditylenchus dipsaci* in the form of "eelworm wool" in dried plant material, have been shown to survive for at least 23 years; however, it is the second-stage larvae of *Anguina tritici* and adults of *Aphelenchoides ritzemabosi* that can resist dehydration. The ecological position of the species is no guide to desiccation resistance; among the soil-inhabiting stages of *D. triformis, D. dipsaci, A. ritzemabosi, Pratylenchus penetrans* and *Tylenchorhynchus claytoni*, the first three were able to survive dry conditions which killed the other two.

Not only the stage but also its state may govern desiccation resistance. Ellenby (1969) has shown that hatched larvae of *Heterodera rostochiensis* can survive low relative humidities for a matter of minutes, whereas eggs in cysts remain viable when stored in dry vegetable products for at least five years. Although water uptake by hatched and unhatched *Heterodera* was the same, illustrating the permeable nature of the egg shell, the rate of water loss was much more rapid from the free larvae after an initial decrease (figure 9.5). The final water content of both the free-living larvae that died

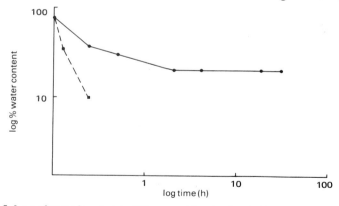

Figure 9.5. Loss of water from hatched ■ – – – – –■ and unhatched ●————● *Heterodera rostochiensis* larvae (redrawn from Ellenby, 1969).

and those within the egg which survived, was the same. It was thus suggested that the slowed rate of drying due to the structure of the egg shell had enhanced survival. As the eggs are normally retained within a tanned protein cyst which represents the remains of the female's body, the rate of drying is probably even slower than in free eggs. Any moisture remaining within an impermeable cyst will tend to keep the relative humidity high, and even further slow the rate of drying of the enclosed eggs. Support for the theory that the rate of drying may be influential in desiccation survival has also been provided by Ellenby (1969). Ensheathed third-stage larvae of *H. contortus* were shown to survive appreciably longer than exsheathed larvae, and their rate of water loss was also slower (figure 9.6, 9.7). By using the interference microscope it was possible to show that the sheath itself dried more rapidly than the enclosed larva, and in the dried state provided an effective barrier to further water loss. Survival of *D. dipsaci* in "eelworm wool" is also apparently governed by similar factors. In small quantities of wool, all the nematodes survive equally but, in larger masses, survival of the larvae in the centre, and thus protected from the greatest effects of dehydration, was much greater than for those at the periphery.

It is not possible to explain all the phenomena of desiccation resistance in terms of rate of drying. Little experimental support has been found for theories that differences in lipid content, or in the distribution of the water in the different species, may play a part. The basic reasons why some nematodes are better able to survive than others, remain obscure.

Bhatt and Rohde (1970) have investigated the metabolic rate of *Anguina tritici* recovering from a state of desiccation. They found no detectable oxygen uptake in desiccated larvae, but they did report that oxygen uptake was resumed when they were placed in water. Maximum uptake was achieved in 6–8 hours by individuals desiccated for one year, but it took 70–80 hours to achieve the same uptake after 10 years of dehydration. From these results, it appears possible that during desiccation gradual physiological changes take place that take longer to reverse as the period of drying is increased. It may be the ability of the nematode to reverse these changes that governs its ability to resist desiccation.

The importance to a given species of the ability to resist extremes of desiccation is debateable. Although these conditions can be imposed almost indefinitely in the laboratory, they are very unlikely to persist for more than a matter of days in the micro-habitats in which most nematodes live.

The effect of reduced temperature on nematodes (like that of dehydration) varies widely with the species concerned. Within the general range 10-35°C, activity and metabolic rate follow the temperature. Crop damage has often been related to soil temperature, and reduction in motility may greatly influence the level of infection, which is generally an active process.

Although many nematodes recover from being cooled to 5°C, below this temperature different species behave differently; some may be regarded as resistant, and others as susceptible to freezing. Free-living larvae of the

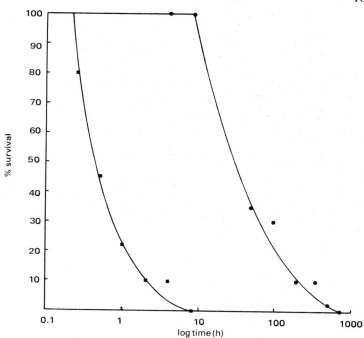

Figure 9.6. Survival of ensheathed ●————● and exsheathed ■————■ *Haemonchus contortus* third-stage larvae (redrawn from Ellenby, 1969).

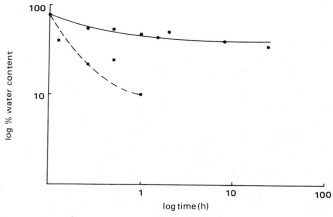

Figure 9.7. Loss of water from ensheathed ●————● and exsheathed ■ – – – ■ *Hae-monchus contortus* larvae (redrawn from Ellenby, 1969).

animal parasites *Strongylus edentatus*, *S. vulgaris* and *Nematodirus filicollis* can survive freezing for up to a year, but hookworm larvae and those of *Angiostrongylus cantonensis* are killed by freezing. These larval stages are more resistant than the susceptible adults which, as parasites of mammals, require a closely controlled temperature range. Among the plant parasites, *A. avenae* will resist freezing, and fourth-stage larvae of *P. penetrans* can recover after cooling to –4°C. Species such as *Trichodorus christiei* and *M. javanica* do not fully recover from 3°C. Extremes of temperature have been used on some nematodes: *D. dipsaci* fourth-stage larvae have been cooled to –80°C, freeze dried and stored for five years, and a species of *Plectus* stored for 125 hours at –190°C; both species recovered from the treatment. The factors governing the ability of an organism to resist freezing temperatures are not fully understood. It has been suggested that the chemical composition of the lipids forming the nuclear membrane determines the ability to resist chilling between 10° and 0°C. Cytoplasmic composition may also be responsible, as this may influence the size of the ice crystals formed on freezing, and thus affect the degree of cellular damage. These factors have not been measured for nematodes and their influence must remain uncertain.

The third environmental condition that may induce dormancy in nematodes is anaerobiosis. The oxygen content of the habitats of nematodes varies widely from the almost anaerobic centre of the lumen of the vertebrate colon or marine muds, to the aerobic insect haemocoel, or the surface of aerial plant tissues. The oxygen level may also vary in any given habitat; thus in wet heavy soils, more or less anaerobic conditions may be encountered, whereas in light sandy soils the soil oxygen level approximates to that of the air. The ability of given species of nematodes to overcome periods of anaerobiosis almost certainly influences their natural habitat.

The initial response of nematodes to reduced oxygen appears to be a loss of motility (Table 9.3). The oxygen stress for these results was produced by the oxidation of sodium sulphite to sodium sulphate by atmospheric oxygen, and the actual oxygen concentration was not measured. It is apparent that a number of nematodes can withstand reduced oxygen levels for an appreciable length of time. Aerobically, nematodes utilize an oxidative metabolism with neutral lipid catabolism as the main energy-producing process. Under anaerobic conditions there may be a change to fermentative metabolism, the major metabolite being glycogen. The ability of the nematode to make this change may govern its survival under anaerobic conditions. Starved *Caenorhabditis* spp. survived anaerobically for only 80

Table 9.3. Influence of reduced oxygen tension on the motility of various nematode species (Feder and Feldmesser, 1955)

Species	Time till cessation of movement (min)	Recovery time (min)
Aphelenchoides obsistus	4	2
Radopholus similis	7–9	5–10
Belonolaimus gracilis	10	Immediate
Tylenchulus semipenetrans	10–12	10
Rhabditis sp.	15	2
Heterodera rostochiensis larvae	15	2
Meloidogyne sp. larvae	20	2
Dolichodorus heterocephalus	45	5–10

hours, whereas *A. avenae* remained alive for at least 30 days (Cooper and Van Gundy, 1971). *Caenorhabditis* used some of their stored glycogen under aerobic conditions and progressively more as the oxygen content of the environment was reduced. By 72 hours all detectable glycogen had been utilized. Aerobically *A. avenae* used no glycogen, and there was even some remaining after 20 days of anaerobiosis.

CHAPTER TEN

TREATMENT AND PREVENTION OF
NEMATODE DISEASES

A MAJOR FORCE BEHIND THE SEARCH FOR NEW KNOWLEDGE ABOUT
nematodes is the desire to control or eliminate those species which cause
disease in man, his animals and his crops. Some of the methods used
include crop rotation, the selection of resistant varieties of host, and simple
sanitation. These forms of cultural practice have been used for centuries,
and were almost certainly discovered by simple trial and error. The Jewish
habit of avoiding pork prevents exposure to *Trichinella spiralis,* and
Buddhist monks who filter all the water they drink (to be sure that they
eat no living creatures) also filter off many infective stages of parasites.

Although well known in Europe, the first recorded outbreak of human
trichinosis in tropical Africa was in 1961, when 11 youths killed a wild bush
pig on the slopes of Mt. Kenya and ate the undercooked flesh. These were
also the heaviest trichinosis infections ever recorded in man (Nelson, 1972).
T. spiralis is known to occur in East Africa in the bush pig, the leopard, the
jackal, the serval, the lion, the spotted and striped hyaena and the domestic
dog. The outbreak in Kenya occurred as a result of a change in human
behaviour. The traditional Kikuyu custom was to refuse uncooked flesh
but, after the Mau-Mau insurrection, many customs such as this were
abandoned. The youths were therefore infected; no women or small
children were infected, because they continued to eat only well-cooked
meat.

The removal of the gravid female of the Guinea worm *Dracunculus
medinensis* from the skin is performed by winding it onto a stick or some
similar object. It is believed that this is part of the origin of the symbol of
the medical profession—a snake coiled around a staff (figure 10.1). When
Western scientists first started to investigate onchocerciasis or tropical
river blindness in Central Africa, they were taken to the bank of a river and
told by the native peoples that the disease came from the river. Later, the
blackfly *Simulium* spp., which breeds only in fast-flowing streams, was
demonstrated to be the intermediate host of *Onchocerca volvulus.* Early

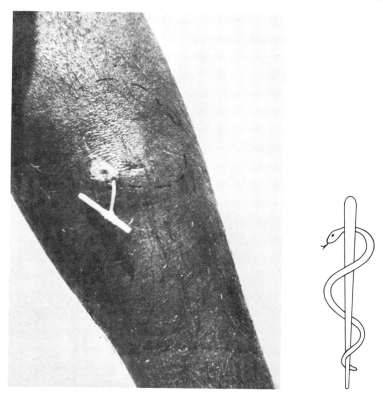

Figure 10.1. *Dracunculus medinensis* wound around a forked stick and the emblem of the medical profession which may have been derived from it. (Photograph by courtesy of the Wellcome Foundation).

drugs were derived from ferns, oils and various "medicinal" plants. These traditional sources are now taken very seriously by pharmaceutical companies who are trying to find new drug treatments. The fascinating area of parasitology, the causality of parasites with disease and methods of treatment in pre-science cultures, is now being actively pursued, but is outside the scope of this book.

Chemotherapy

Drugs are the most widely used form of treatment against parasitic infestations. These range from simple tablets, available from the household chemist, for "worming" children and pets, to the highly dangerous organophosphorus and cyanide derivatives which are used against some veterinary and human filarial infections. Drugs used against the nematode parasites of man and other animals are called *anthelminthics* (or anthelmintics), while *nematicides* is the name used for drugs against plant-parasitic nematodes. The distinction is, of course, entirely artificial and has no more basis in science than the division between helminthology and nematology.

Some of the drugs, like chenopodium oil, which contains ascaridole, a broad-spectrum anthelminthic, were developed from traditional remedies. Other drugs with activity against nematodes, such as the organophosphorus derivatives were first used in insecticides, but were later tested against nematodes. *Piperazine,* a most successful anthelminthic was first administered clinically at the turn of the century as a uriscosuric agent for the treatment of gout. Its low mammalian toxicity and clear anthelminthic properties were discovered accidently, and it was first used for the control of worms in 1949. Pharmaceutical firms, with their great preoccupation with cost-effectiveness, have repeatedly found that most new drug formulations against nematodes have resulted from an empirical search amongst numerous possible compounds. Once a likely compound is selected, then it is screened *in vitro* and *in vivo,* and chemical elaborations by chemists enable a family of related compounds to be tested. The testing of drugs *in vivo* has been much more productive

Table 10.1. Examples of nematodes used in primary and secondary screening of chemotherapeutic drugs.

Disease	Primary screen	Secondary screen
Ancylostomiasis (hookworm)	*Nematospiroides dubius* (mouse)	*Ancylostoma caninum* (dog)
	Nippostrongylus brasiliensis (rat)	*Uncinaria stenocephala* (dog)
Strongyloidiasis	*Strongyloides ratti* (rat)	*Strongyloides papillosus* (sheep)
Trichuriasis	*Trichuris muris* (mouse)	*Trichuris vulpis* (dog)
Filariasis	*Litomosoides carinii* (cotton rat)	*Brugia malayi* (cat)
Parasitic bronchopneumonia	*Dictyocaulus viviparus* (guinea pig)	*D. viviparus* (cattle)
	D.filaria (guinea pig)	*D.filaria* (sheep)

than using *in vitro* methods. Of the modern anthelminthics, *bephenium,
diethylcarbamazine, dithiazanine, piperazine* and *pyrvinium* have all
emerged through *in vivo* screens. *In vitro* testing has produced nothing
of practical value yet (Table 10.1). Unfortunately, there is insufficient
knowledge of nematode biochemistry and physiology to enable a more
rational approach to the development of drugs. In a few cases the mode
of action of the drugs is understood, and this information is teaching us
some basic nematology. This information has come in every case after
the application of the drug, which underlines the biological ignorance of
formulation.

Evans (1973) has emphasized that there are many steps between the

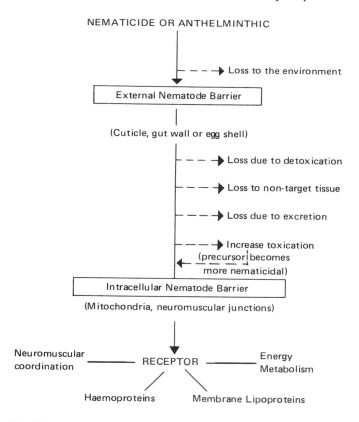

Figure 10.2. Flow diagram representing the penetration of a nematicide to a receptor site
(modified from Evans, 1973).

application of a drug and its effect on the nematode. This is particularly useful, as it is too often assumed that there are only two stages : the application of the drug and the death of the nematode. In figure 10.2 it may be seen that there is a considerable loss of drug to the environment. Following this there is often the problem of penetration. Some nematodes have been found to survive for days in extremely high doses of drugs without any effect, but if they are induced to feed, they quickly die. The form in which the drug is presented can be critical. The hope of formulating an "ideal" nematicide for plant-parasitic forms is to find a downwardly translocated systemic compound, with no phytotoxicity. This would be applied to the aerial parts of plants and transported to the roots, where it would enter those nematodes feeding on cell sap.

Once inside the tissues, there is a loss of drug to non-target tissues, a detoxication of the drug may occur, and the drug will be excreted. In some drugs there is a metabolic action to produce products even more toxic than the original parent compound. After crossing cellular and intracellular barriers, the drug acts against a "target" or, to conform with pesticide nomenclature, a "receptor". There are four major "receptor" sites in nematodes: neuromuscular coordination, haemoproteins, energy metabolism and membrane lipoproteins.

Piperazine is a most successful broad-spectrum drug which is widely used against *Ascaris lumbricoides* in man and pigs and *Enterobius vermicularis* in man. Six years after its introduction as an anthelminthic, it was found that *A. lumbricoides* were paralysed but not dead when expelled following administration of the drug. If they were immersed in warm isotonic saline, they soon recovered and swam actively. This observation suggested that there was a reversible action on the neuromuscular system. The immobilized worms are evacuated by the host's gut movement. Since that time the mode of action has been thoroughly studied, using nerve-muscle preparations of *A. lumbricoides* (reviewed by del Castillo and Morales, 1969). The excitatory action of acetylcholine on *A. lumbricoides* nerve-muscle preparations is blocked by piperazine. Piperazine increases the membrane potential of muscle cells, causing hyperpolarization which antagonizes the depolarizations induced by acetylcholine.

The narcotizing action of piperazine is effective because the host expels the worms. Organophosphates such as *dichlorovos* which began as an insecticide, are broad-spectrum anthelminthics which like carbamates and pyrvinium pamoate act on acetyl cholinesterase to disrupt neuromuscular co-ordination. These compounds all narcotize nematodes and their effects are ususally reversible (figure 10.3).

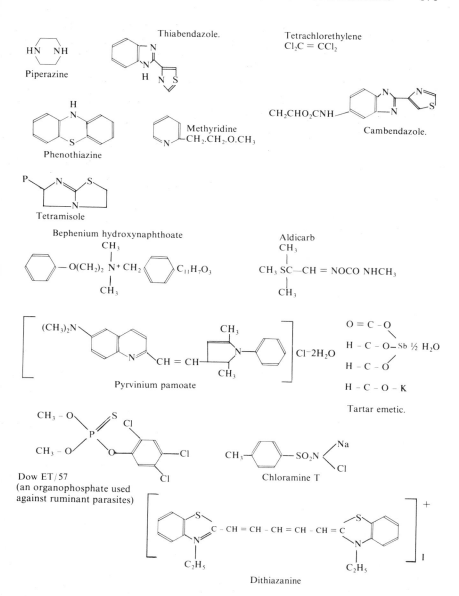

Figure 10.3. Structural formulae of some anthelminthics and nematicides.

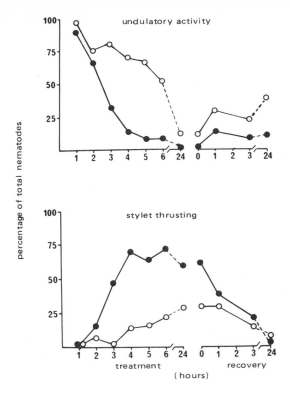

Figure 10.4. The movement of second-stage larvae of *Heterodera rostochiensis* in the nematicide Aldicarb.
●————● 1 part per million
o————o 10 parts per million (adapted after Nelmes, 1972).

Nelmes (1972) examined the effects of the carbamate, *aldicarb,* and found that the second-stage larvae of *Heterodera rostochiensis* were immobilized. The larvae started spontaneous movements of the stylet when inactive (figure 10.4). Similar reversible uncoordinated movements have been observed in other anti-cholinesterase compounds. The concept of the "nematostat" is now being added to that of the anthelminthic and the nematicide. The "nematostat" is a reversible narcotizing agent which usually acts on neuromuscular co-ordination and is a non-lethal paralysant which kills only at high doses and after long exposures. There are close parallels between nematicide-induced narcosis and normal inactivity or quiescence. Many nematodes are induced to enter a quiescent state of reduced physiology through desiccation, cold, or low oxygen concentration. Narcosis alone is therefore not likely to cause death in these species.

The most effective drug for the chemotherapy of human filariasis is *diethylcarbamazine,* one of the piperazine series, discovered in 1947. It illustrates another of the basic phenomena in the practice of chemical control against nematodes.

Diethylcarbamazine is primarily active against microfilariae, and less effective against adults. A single intravenous injection into a cotton rat infected with *Litomosoides carinii* causes 80% of the microfilariae to disappear from the peripheral circulation within 2 minutes. A similar clearing action occurs in *Wuchereria bancrofti* and *Brugia malayi* infections following an oral dose of the drug. After the treatment, the microfilarial stages accumulate in the small vessels of the liver. Macrophages congregate about the entrapped microfilariae and destroy them within 20 hours, presumably by phagocytosis. This is an example of a drug which is primarily active against only one stage of the parasite. Each larval stage and the adult is a separate physiological entity, and is susceptible to only a limited range of drugs.

In the treatment of ruminants, it has been found that the therapeutic dose of organophosphates is often near the level of mammalian toxicity. Organophosphates, which are, of course, related to the well-known insecticide DDT, are of limited usefulness in the treatment of mammals. *Haloxon* is something of an exception, and the minimum oral dose is stated to be 250mg/kg, which is some five times less than the lethal dose.

Antimonials, arsenicals and cyanide dyes have all been used in the treatment of hosts infected with filarial nematodes. Lithium antimonythiomalate, neostibosan and antimony potassium tartrate (tartar emetic) are all quite effective, but their general usage is restricted because of their severe human toxicity. The most effective arsenical is *Mel W* (pentylthiarsuphenylmelamine) which is only mildly toxic to man. A non-toxic chemotherapeutic agent active against both microfilariae and adult filarids is urgently required (Desowitz, 1971). *Dithiazanine* is one of the few compounds other than piperazine about which the mode of action is reasonably understood. Dithiazanine is a cyanide dye with a broad antinematode spectrum. The chief use of "dyes" of this type is as colour sensitizers in photography. *Litomosoides carinii* is an obligate aerobe that lives in the pleural cavity of cotton rats. Low concentrations (6.5×10^{-6} M) of a number of cyanide dyes inhibit oxygen consumption by adult *L. carinii.* Associated with the respiratory inhibition produced by the cyanides is an increased rate of aerobic glycolysis, and a decreased rate of polysaccharide synthesis. The dyes are therefore believed to exert their chemotherapeutic effect by inhibiting an enzyme system associated with oxidative metabolism.

Dithiazanine is also effective against *A. lumbricoides, Enterobius vermicularis, Strongyloides stercoralis* and the intractable *Trichuris trichiura. Trichuris* is a facultative aerobe. It survives significantly longer outside the host in 5% CO_2 in nitrogen than in 5% CO_2 in air or oxygen. The drug inhibits the uptake of glucose by *T. vulpis*. This inhibition is demonstrable at a drug concentration lower than that causing immobility of the parasite. When kept in dithiazanine for 24 hours and then transferred to drug-free medium, the inhibition of glucose uptake is maintained or increased. Therefore, unlike piperazine, the effects of dithiazanine are irreversible. The depletion in the utilization of glucose is related to a reduction in the amount of glucose in the tissues. This has been interpreted to mean that it is not the enzymes which metabolize glucose that are inhibited, but rather the uptake of glucose itself. Dithiazanine has no effect on hexokinase activity of *T. vulpis*, further indicating that there is no effect on intracellular utilization. ATP and glycogen levels are lowered by the drug. On the basis of these findings it has been concluded that dithiazanine reduces the transport of exogenous glucose in *T. vulpis*. Dithiazanine is unfortunately fairly toxic to humans, and there have been several deaths associated with abnormal drug absorption.

When dogs infected with the filarial worm *Dirofilaria immitis* are treated with *caparsolate sodium,* the nematode's intestinal epithelium undergoes a series of morphological changes. It has been proposed that the primary site of action of this arsenical drug is on the bacillary border of the intestine, on which an arsenical protein complex serves as an absorption barrier (Lee and Miller, 1969). The worms will therefore eventually starve and die. This may explain the long-term treatment required with this organic arsenical and its lack of effect against microfilariae which have no intestines.

Methyridine or "Promintic" is a most unusual anthelminthic, both in its chemical properties and in the mode of its application. It is a colourless sweet-smelling liquid which is completely miscible with water. A further unusual feature is that it is more effective against parasites of the gut via subcutaneous injection that if administered orally. After injection of the host, methyridine is present throughout the host's tissues very quickly; it then becomes metabolized and excreted. Its metabolites include α picolinic acid, formic acid and formaldehyde, and it was thought that these may be toxic to nematodes. *In vitro* testing, however, demonstrated that the parent compound, methyridine was at least 100 times more nematicidal than any of its metabolites (Broome, 1961).

There is a close association between the levels of methyridine in the host blood and in all parts of the alimentary canal (except the stomach and abomasum of sheep). It is thought that the drug enters the blood, and that an equilibrium is then maintained between its concentration in the blood and all parts of the alimentary tract. It has been shown that the drug readily enters the blood or gut, depending on the concentration gradients. This mode of application is apparently unique amongst known anthelminthics, and it explains why the drug is so useful against a wide spectrum of nematodes at different stations along the alimentary tract. Furthermore, as it is passing back and forth across the mucosa, it is active against species which are buried deeply in the mucosa.

Methyridine immobilizes *A. lumbricoides in vitro* but does not cause any reduction in the rate of oxygen consumption, so the worms are not killed. When the drug is placed on nerve-muscle preparations of *A. lumbricoides,* it induces a contracted paralysis at 10 μg/mg; this effect is not antagonized by acetylcholine.

Thiabendazole is the most useful and widespread anthelminthic employed in the treatment of *A. lumbricoides* and most nematodes which cause parasitic gastroenteritis, as well as against flukes and tapeworms. Fumarate reductase (NADH: fumarate oxidoreductase) is an essential step in the fermentative pathway of *A. lumbricoides.* In the very low oxygen concentrations of the vertebrate gut, *A. lumbricoides* excretes large quantities of succinic, proprionic and other organic acids. The fumarate reductase is probably important for the regeneration of NAD from NADH. This step also provides ATP anaerobically in the fermentation of glucose by *A. lumbricoides.* Prichard (1970) studied the enzyme kinetics of fumarate reductase from homogenates of *A. lumbricoides,* and he concluded that thiabendazole at 10^{-3}M inhibited its activity. A similar "receptor site" has been suggested for *tetramisole.* This is an example of a chemotherapeutic compound which acts against the energy metabolism of the parasite. The metabolic step is critical for the survival of the parasite, but not for the host, and so these are ideal anthelminthics with minimal side effects.

Antiminth, or *pyrantel pamoate* is a very interesting drug of a very new family of chemicals. It is unrelated to currently available medications and is used for the treatment of *A. lumbricoides, E. vermicularis, Ancylostoma duodenale* and *Necator americanus.* Antiminth is more than 1000 times more effective than piperazine in terms of the concentration required to produce an effect (Pitts and Migliardi, 1974). This drug is additionally promising because it is not absorbed by the host intestine and is also highly palatable. The mode of action is thought to be against neuromuscular coordination.

Fumigation of fields and seedbeds is the main method of treating soil to remove plant-parasitic nematodes, although this method has not been found effective for parasites of animals. The fumigants are volatile halogenated hydrocarbons, mostly *alkyl halides,* the most widely used being: *D–D* (1,3-dichloropropene–1,2-dichloropropane), methyl bromide, EDB (ethylene dibromide) and *Telone* (1,3-dichloropropene). The reactions of nematodes and insects to alkyl halides are similar. There is a brief period of hyperactivity which leads to a gradual reduction in activity and eventually complete paralysis; in some situations this narcosis may be reversible on the removal of the fumigant.

The susceptibility of *Aphelenchus avenae* to EDB treatment was related to nematicide treatment, and moulting stages are highly susceptible (Evans, 1973).

Alkyl halide nematicides rapidly oxidize iron porphyrins. Inhibition is lethal, since porphyrins are haemoproteins which are found in cytochromes, haemoglobin and microsomal oxidases. There is no direct evidence for this yet, although EDB toxicity to *A. avenae* decreased in anaerobic conditions. Furthermore, alkyl halide treatment causes a reduction in the rate of respiration. Drugs do not influence all species of nematodes equally (Standen, 1963). The effectiveness of a family of *chloramine* compounds was tested against a wide taxonomic range of soil nematodes, including Rhabditida, Enoplida, Tylenchida and Dorylaimida. The Enoplida were consistently most susceptible, then the Dorylaimida and *Rhabditis, Acrobeles* and *Tylenchus* were most resistant to action (Viglierchio and Croll, 1969) (figure 10.5).

One of the problems which is always expected in the use of drug application is the selection of genetically drug-resistant populations. As is well known, bacteria, malaria, mosquitoes and other pests have all developed drug-resistant strains, and a continuous campaign is waged to create new formulations as these pests adapt to the existing ones. This problem has been recognized in some parasitic nematodes, but it is not yet a widespread phenomenon.

Some populations of *Haemonchus contortus* in sheep and cattle and *Trichostrongylus colubriformis* in sheep have developed resistance to thiabendazole, parabendazole and cambendazole (i.e. benzimidazole compounds). Other populations of *H. contortus* are resistant to phenothiazine

1% Trichloromelamine

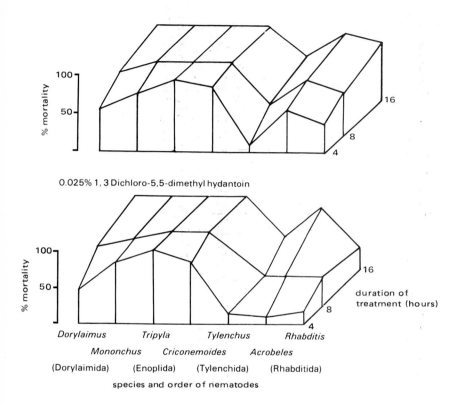

0.025% 1, 3 Dichloro-5,5-dimethyl hydantoin

Figure 10.5. A comparison of the activity of soil nematodes following treatment in various chloramines: *(a)* 1% Trichloromelamine *(b)* 0.025 1,3–dichloro–5,5 dimethylhydantoin (from Viglierchio and Croll, 1969). (E.J.Brill, Leiden)

and *Ostertagia circumcincta* in sheep has become resistant to some organophosphorus compounds. Kates, Colglazier and Enzie (1973) exposed *H. contortus* in four successive generations of lambs to doses of cambendazole and established a partial drug resistance. The surviving parasites were then exposed to further drug exposures at increased dose levels. After the tenth exposure, the magnitude of resistance had increased substantially. Genetic selection of drug-resistant populations of nematodes is therefore possible, and should be incorporated into a wider perspective of anthelminthic chemotherapy.

Vaccination against nematodes

Vaccination and immunization are familiar for such diseases as yellow fever, cholera and smallpox, but the complexity of the immunology of nematodes has hindered development of an effective vaccine. Only one nematode species, *Dictyocaulus viviparus,* is routinely controlled by vaccination in 1976. Studies continue on vaccines for hookworm and other nematodes.

Dictyocaulus viviparus causes parasitic bronchitis or "husk" in cattle; the worms were first reported as long ago as 1744, and their pathogenicity was realized in 1756. The vaccine was developed in the late 1950s and early 1960s, largely by Jarrett, Poynter and their respective collaborators (Poynter, 1963). *D. viviparus* lives in the lungs of cattle and lays eggs which pass to the field, develop to infective third-stage larvae, and enter contaminatively on herbage. Once in the gut, the larvae are believed to penetrate into the mucosa and enter the mesenteric lymph nodes; after a lengthy migration, they arrive at the lungs. It is during this migration that the host produces antibodies against the worm. Injection of larval homogenates does not cause protection, but the use of attenuated live vaccine does.

Infective larvae are X-irradiated at doses of 20,000 to 40,000 r; this weakens them, but they are still able to enter the mesenteries of cattle and calves, and begin their migrations. During their short life they are apparently able to produce antigenic material to which the host manufactures effective antibodies. On or before their arrival at the lungs, the larvae die, and so the host is protected from further infection without suffering the full extent of the disease. This live vaccine has now been used successfully all over the world for the control of husk.

Physical methods of control

Nematodes are killed by ultraviolet and X-irradiation and by electric currents, but there has not been any effective use of these techniques in their direct control. Ultrasonics have also been suggested but have not proved practicable. The excessive application of sucrose on fields contaminated with plant-parasitic nematodes can theoretically control them. Sugar is non-toxic and easily stored. Osmotic control using sugar is not practical; it is currently condemned because 20–30 tons/acre of sucrose is needed to enhance a maximum production of sugar from sugar beets of 4 tons/acre!

"Hot water treatment" is widely used for the control of *Ditylenchus dipsaci* in daffodil and tulip bulbs. The technique depends upon the greater susceptibility of the nematodes than the bulbs to heat. Infected bulbs are immersed in water baths at about 46°C for 8 minutes. This gives good control of the nematode, with a small and tolerable loss of bulbs. The principle of differential host and parasite susceptibility to heat has been applied to other nematodes. For sweet potatoes, "dry heat" of 45°C for 30 hours gave complete control of nematodes.

Control of disease vectors

It would be inappropriate to give a detailed account of mosquito and blackfly control in these pages. Widespread campaigns are mounted against the vectors of nematode diseases which are transmitted by arthropods.

Biological control of nematodes

Biological control has most seriously been considered for the control of soil-dwelling plant-parasitic nematodes. All of these forms spend a large proportion of their lives in the soil, and the rationale has been to exploit the naturally-occurring soil pathogens for control.

Protozoa have been found associated with nematodes but, as Canning (1973) emphasizes, the literature is very confused.

Unfortunately many of these were poorly described: some may not even be protozoans and for those that are, insufficient information was given in many cases to make valid diagnoses of genera. Probably many were incorrectly assigned at the ordinal level (p.342).

Many of the Sporozoa transmitted to nematodes enter in the food as resistant spores or cysts, e.g. *Legerella helminthosporum* (Sporozoa, Coccidia). When the spores are ingested, the cyst is broken down, and sporozoites invade the cells. The free-living nematode *Mononchus composticola* can readily be infected; however, plant parasites have a stylet with a canal which is too small to permit the passage of spores or bacteria. The gregarines and coccidians can, therefore, be excluded as suitable candidates for the biological control of plant-parasitic nematodes.

Some flagellates are found in association with nematodes, e.g. *Bodo caudatus* (Kinetoplastida), a common soil form, has been found on cultures of *Ditylenchus dipsaci*. Dense aggregations were found around the excretory pore (figure 10.6), where they were probably feeding upon bacteria.

At least one unicellular parasite is known to penetrate the cuticle of *Meloidogyne* spp., *Pratylenchus* spp. and *Aphelenchoides* spp., and to

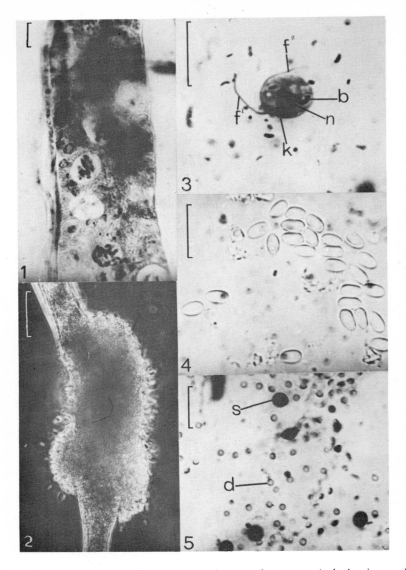

Figure 10.6. (1) Longitudinal section of part of *Mononchus composticola* showing nearly mature oocysts of *Legerella helminthosporum* in the cells of the gut. (2) Aggregates of *Bodo caudatus* and bacteria around the excretory pore of *Ditylenchus dipsaci*. (3) *Bodo caudatus* n, nucleus; k, kinetoplast; f, flagella; b, bacterium in food vacuole. (4) Fresh preparation of microsporidian spores (*Nosema eurytremae* from trematode larvae, as an example to show general appearance). (5) Spores of *Duboscquia penetrans* in a section of *Meloidogyne javanica*. s, strongly Gram-positive spores; d, dome of spore wall showing Gram-positive rim. Scales 1,3,4,5 = 10μm, 2 = 50μm (Canning, 1973).

cause the death of these nematodes. As these genera include some extremely serious agricultural pests, the potential usefulness of this parasite cannot be underestimated. Regrettably, the status of this parasite is unclear. *Duboscquia* is the name given to a genus of microsporidian protozoans, but there is a serious doubt regarding the identity of the parasite called *D. penetrans* on nematodes; it may even be a fungus. The "spore" clamps onto the cuticle and somehow enters the pseudocoelom of the nematode. The gonads and other tissues are destroyed, leaving the body packed with many hundreds of spores.

On many unrelated occasions, it has been found that organic additives reduce the numbers of plant-parasitic nematodes. Such things as green leaf mould, castor pomace (a byproduct of castor oil production) and manure have all been shown to reduce the numbers of nematode pests under certain conditions. In these cases it has also been found that the density of natural parasites and predators has been increased. Enchytraeids, tardigrades, collembolans, amoebae and ciliates have all been shown to reduce the numbers of nematodes by direct predation. Of these, only two groups of organisms have been seriously investigated.

The first of these groups is the predacious or nematode-trapping fungi (Duddington, 1956). These fungi trap nematodes in the soil on sticky surfaces or in nooses. One of the "sticking" fungi is *Arthrobotrys oligospora;* the hyphae grow into an anastomosing network of sticky surfaces into which nematodes move. They are quickly trapped and hyphae penetrate the cuticle within two hours; small trophic hyphae soon fill the nematode, digesting and absorbing its contents. There is a number of other fungal genera which also use sticky secretions.

The nematode trapping forms have little rings made of three adjacent cells which are attached to the parent hypha by a short stalk. In *Dactylaria candida* the dimensions of the noose are critical, because small nematodes or larval stages can slip directly through it. In other species, rapid turgor changes occur in the hyphal cells; they increase their volume by three times, and the noose constricts around the body. In both species trophic hyphae rapidly penetrate the cuticle to digest and absorb the body contents.

There have been a few detailed examinations of nematode-trapping fungi, and even some field trials. Unfortunately the results are not encouraging. The fungi are widespread in nature and probably take their toll of nematodes, even under diseased crops. The soil conditions appear to be critical for survival of the fungi and other natural pathogens and predators of nematodes; and the method of application, the speed of control and the lack of specificity are all unsatisfactory at present.

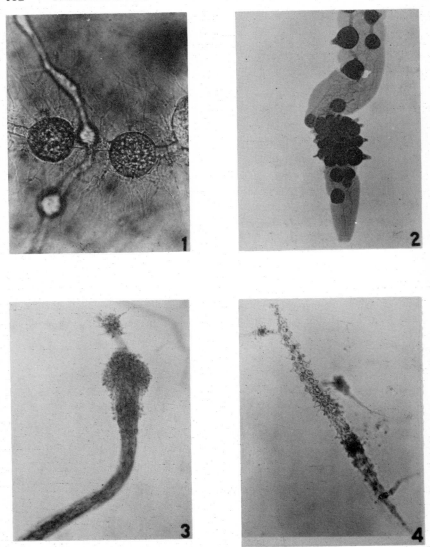

Figure 10.7. Photomicrographs of *Catenaria anguillulae* showing its hyphae, zoosporangia and rhizoids growing (1) on an agar medium and (2) in the nematode *Panagrellus redivivus*, (3) Zoospores accumulating at the mouth and excretory pore of *P. redivivus*. (4) The carcus of *P. redivivus* completely filled with zoosporangia (Sayre and Keeley, 1969).

Although they are biologically intriguing, nematode-trapping fungi hover in the background until economic or biological pressures are sufficient to re-consider their possible usefulness (figure 10.7).

Another potentially useful fungus, the phycomycete *Catenaria anguillulae,* has been tested for its suitability. In sand or soil, the efficiency of this predatory fungus to control plant-parasitic nematodes and its lack of specificity were not sufficient to recommend it over existing methods (Sayre and Keeley, 1969).

A similar situation exists for the predatory nematodes. There are a few groups of highly voracious nematodes which feed on other species of their own phylum. Many of them are in the Mononchidae of the Dorylaimida, but *Seinura* is a stylet-bearing form in the Aphelenchoididae and *Mononchoides* is in the Diplogasteridae. Predatory nematodes prey upon many types of soil animals in addition to nematodes, and will feed on insects, tardigrades, protozoans, oligochaete annelids, and rotifers. When a large individual of *Mononchus, Prionchulus* or *Iotonchus* feeds on a small nematode, it places its lips against the cuticle and, with the powerful suction produced by its pharynx, it tears through the body wall of the prey. The stoma of such species has special hooks or "onchia" in the stoma, and these shred the body as it passes rapidly into the intestine.

The "Father of Modern Nematology", Nathan Augustus Cobb (1859–1932) described in vivid terms the feeding habits of *Mononchus.*

Obviously mononchs hunt by the aid of some sense other than sight, since both they and their prey usually live in subterranean darkness. Picture these ferocious little mononchs engaged in a ruthless chase in the midst of stygian darkness. We may imagine them taking up the scent of the various small animals upon which they feed, among which almost anything they can lay mouth to seems not to come amiss, and pursuing them with relentless zeal that knows no limit but repletion (Cobb, 1917).

(Research has confirmed Cobb's description of voracity, but not all his speculations about olfactory location of prey.)

In some predatory nematodes it has been found that there is no specificity, and they will consume all species offered (Yeates, 1969). In *Seinura,* cannibalism has been observed between predators, but this tends to be under conditions of poor nutrition, and the individuals are often moribund. The number of prey consumed by each one of these predators may be tens or even hundreds in a few days, and so the potential for biological control is clear. Biological materials are, however, delicate agents to apply, and require a refinement of manipulation not possible with current agricultural methods. Like protozoans and nematode-trapping fungi, predatory nematodes are biologically dramatic and have aroused interest but must remain only theoretical possibilities at the present.

Integrated control of nematodes

Crop rotations, resistant varieties, natural predators and parasites, and physical methods are not readily adaptable to medical parasitology. In medical practice, sanitation and chemotherapy remain the primary control methods.

Biological or chemical control methods are not now seen as alternatives; instead, the contemporary attitude is to establish an "integrated control". This assembles all the measures available and uses them in various combinations to the maximum benefit. In this context, for example, the result of fumigant nematicides has been viewed in respect of the natural populations of predators in the soil. It has been observed many times that, following application of chemicals to crops, and obtaining short-term control of the nematode pests and good crop yields, even more chemical is needed for the same yield in the following season. It is very hard to monitor this kind of effect exactly, but it has been strongly argued that the chemical has not only suppressed the target nematode pest, but also killed the natural predators and parasites of the pest. Perhaps one of the real contributions of the studies into biological control has been to authenticate the hazards to "beneficial organisms" which can be caused by chemicals.

It is the judicious and well-informed use of crop rotation, sanitation, resistant varieties of hosts, organic additives, vaccines, physical treatments, chemotherapy and biological agents which now characterizes effective control of those nematodes that challenge man's health and agriculture.

FURTHER READING

Chapter 1

Bird, A. F. (1971) *The Structure of Nematodes*, Academic Press, London and New York.

Chitwood, B. G. and Chitwood, M. B. (1950) *An Introduction to Nematology*, Monumental Printing Co., Baltimore, Maryland.

Crofton, H. D. (1966) *Nematodes*, Hutchinson, London.

Hyman, L. H. (1951) *The Invertebrates: III Acanthocephala, Aschelminthes and Entoprocta. The Pseudocoelomate Bilateralia*, McGraw-Hill, New York, London and Toronto.

Lee, D. L. (1965) *The Physiology of Nematodes*, Oliver and Boyd, Edinburgh and London.

Remane, A. (1963) "The Systematic Position and Phylogeny of the Pseudocoelomates" in *The Lower Metazoa—Comparative Biology and Phylogeny* (ed. Dougherty, E. C.), University of California Press, Berkeley and Los Angeles.

Wallace, H. R. (1963) *The Biology of Plant Parasitic Nematodes*, Edward Arnold, London.

Chapter 2

Beguet, B. (1972) "The Persistence of Processes Regulating the Level of Reproduction in the Hermaphrodite Nematode *Caenorhabditis elegans,* despite the influence of Parental Ageing, over Several Successive Generations," *Experimental Gerontology,* **7,** 207–218.

Beguet, B. and Brun, J. L. (1972) "Influence of Parental Ageing on the Reproduction of the F_1 Generation in a Hermaphrodite Nematode, *Caenorhabditis elegans,*" *Experimental Gerontology,* **7,** 195–206.

Brenner, S. (1974) "New Directions in Molecular Biology," *Nature.* London, **248,** 785–787.

Cooper, A. F. Jr. and Van Gundy, S. D. (1970) "Metabolism of Glycogen and Neutral Lipids by *Aphelenchus avenae* and *Caenorhabditis* sp. in Aerobic, Microaerobic, and Anaerobic Environments," *Journal of Nematology,* **2,** 305–315.

Epstein, J., Himmelhoch, S. and Gershon, D. (1972) "Studies on Ageing in Nematodes III. Electronmicroscopical Studies on Age-associated Cellular Damage," *Mechanisms of Ageing and Development,* **1,** 245–255.

Evans, A.A.F. (1970) "Mass Culture of Mycophagous Nematodes" *Journal of Nematology,* **2,** 99–100.

Fiakpui, E. Z. (1967) "Some Effects of Piperazine and Methyridine on the Free-living Nematode *Caenorhabditis briggsae* (Rhabditidae)," *Nematologica,* **13,** 241-255.

Gershon, D. (1970) "Studies on Ageing in Nematodes I. The Nematode as a Model Organism for Ageing Research," *Experimental Gerontology,* **5,** 7–12.

Jenkins, D. C. and Phillipson, R. F. (1972). "Increased Establishment and Longevity of *Nippostrongylus brasiliensis* in Immune Rats Given Repeated Small Infections," *International Journal for Parasitology,* **2,** 105–111.

Kisiel, M., Nelson, B., Zuckerman, B.M. (1972) "Effects of DNA Synthesis Inhibitors on *Caenorhabditis briggsae* and *Turbatrix aceti,*" *Nematologica,* **18,** 373–384.

Peters, B. G. (1952) "Toxicity Tests with the Vinegar Eelworm I. Counting and Culturing," *Journal of Helminthology,* **26,** 97–110.

Petrus, Borellus (1956) *Observationium Microscopicarum Centuria,* Hagae-Comitum.
Samoiloff, M. R. (1973) "Nematode Morphogenesis: Localization of Controlling Regions by Laser Microbeam Surgery," *Science,* **180,** 976–7.
Sayre, F. W., Hansen, E. L. and Yarwood, E. A. (1963) "Biochemical Aspects of the Nutrition of *Caenorhabditis briggsae,*" *Experimental Parasitology,* **13,** 98–107.
Vanfleteren, J. R. and Roets, D. E. (1972) "The Influence of Some Anthelminthic Drugs on the Population Growth of the Free-living Nematode, *Caenorhabditis briggsae* and *Turbatrix aceti* (Nematoda: Rhabditida)." *Nematologica,* **18,** 352–338.
Ward, S. (1973) "Chemotaxis by the Nematode *Caenorhabditis elegans*: Identification of Attractants and Analysis of the Response by Uses of Mutants," *Proceedings of the National Academy of Science,* **70,** 817–821.
Zuckerman, B. M., Himmelhoch, S., Nelson, B., Epstein, B., Kisiel, M. (1971) "Ageing in *Caenorhabditis briggsae*", *Nematologica,* **17,** 478–487.
Zuckerman, B. M., Nelson, B., Kisiel, M. (1972) "Specific Gravity Increase of *Caenorhabditis briggsae* with Age," *Journal of Nematology,* **4,** 261–262.
Zuckerman, B. M. Himmelhoch, S., and Kisiel, M. (1973) "Fine Structure Changes in the Cuticle of Adult *Caenorhabditis briggsae* with Age," *Nematologica,* **19,** 109–112.

Chapter 3
Anya, A. O. (1973) "The Distribution and Possible Neuropharmacological Significance of Serotonin (5-hydroxytryptamine) in *Aspiculuris tetraptera* (Nematoda)," *Comparative and General Pharmacology,* **4,** 149–156.
Bird, A. F. (1971) *The Structure of Nematodes,* Academic Press, New York and London.
Croll, N. A. (1972) "The Behavioural Activities of Nematodes," *Helminthological Abstracts,* **41** (Ser. A) 359–377.
Croll, N. A., Riding, I. L. and Smith, J. M. (1972) "A Nematode Photoreceptor," *Comparative Biochemistry and Physiology,* **40,** A, 999–1009.
del Castillo, J. and Morales, T. (1969) "Electrophysiological Experiments in *Ascaris lumbricoides,*" in *Experiments in Physiology and Biochemistry* (ed. Kerkut, G. A.), II, 209–273, Academic Press, New York and London.
Harris, J. E. and Crofton, H. D. (1957) "Structure and Function in the Nematodes—Internal Pressure and Cuticular Structure in *Ascaris,*" *Journal of Experimental Biology,* **34,** 116–130.
Hope, W. D. (1974) *"Deontostoma timmerchioi* n.sp., a New Marine Nematode (Leptosomatidae) from Antarctica, with a note on the Structure and Possible Function of the Ventro-median Supplement," *Transactions of the American Microscopical Society,* **93,** 314–324.
Lee, D. L. (1962) "The Distribution of Esterase Enzymes in *Ascaris lumbricoides,*" *Parasitology,* **52,** 241–260.
McLaren, D. J. (1972) "Ultrastructural and Cytochemical Studies on the Sensory Organelles and Nervous System of *Dipetalonema viteae* (Nematoda: Filarioidea)" *Parasitology,* **65,** 507–542.
Maggenti, A. R. (1964) "Morphology of Somatic Setae: *Thoracostoma californicum* (Nematoda: Enoplidae)" *Proceedings of the Helminthological Society of Washington,* **31,** 159–166.
Rogers, W. P. and Head, R. (1972) "The Effect of the Stimulus for Infection on Hormones in *Haemonchlus contortus,*" *Comparative and General Pharmacology,* **3,** 6–10.
Smith, J. M. (1974) "Ultrastructure of the Hemizonid," *Journal of Nematology,* **6,** 53–55.
Ward, S. (1973) "Chemotaxis by the Nematode *Caenorhabditis elegans:* Identification of Attractants and Analysis of the Response by Use of Mutants," *Proceedings of the National Academy of Science,* U.S.A., **70,** 817–821.
Wright, K. A. and Chan, T. (1973) "Sense Receptors in the Bacillary Band of Trichuroid Nematodes," *Tissue and Cell,* **5,** 373–380.

Chapter 4

Croll, N. A. (1970) *Behaviour of Nematodes,* Edward Arnold, London.

Croll, N. A. (1975) "Behavioural Analysis of Nematode Movement," *Advances in Parasitology,* **13,** 71–122 (ed. B. Dawes) Academic Press, New York and London.

Croll, N. A. and Smith, J. M. (1970) "The Sensitivity and Responses of *Rhabditis* sp. to Peripheral Mechanical Stimulation." *Proceedings of the Helminthological Society of Washington,* **37,** 1–5.

Croll, N. A. and Smith, J. M. (1972) "Mechanism of Thermopositive Behaviour in Larval Hookworms," *Journal of Parasitology,* **58,** 891–896.

Doncaster, C. C. (1971) "Feeding in Plant Parasitic Nematodes: Mechanisms and Behaviour," in *Plant Parasitic Nematodes* (ed. Zuckerman, B. M., Mai, W. F. and Rohde, R. A.), **II,** 137–157, Academic Press, New York and London.

Doncaster, C. C. and Seymour, M. K. (1973) "Exploration and Selection of Penetration Site by Tylenchida," *Nematologica,* **19,** 137–145.

Green, C. D. (1966) "Orientation of Male *Heterodera rostochiensis* Woll. and *H. schachtii* Schm. to their Females," *Annals of Applied Biology,* **58,** 327–339.

Greet, D. N. (1964) "Observations on Sexual Attraction and Copulation in the Nematode *Panagrolaimus rigidus* (Schneider)," *Nature,* London, **204,** 96–97.

Klingler, J. (1970) "The Reaction of *Aphelenchoides fragariae* to Slit-like Micro-openings and to Stomatal Diffusion Gases," *Nematologica,* **16,** 417–422.

Klingler, J. and Kunz, P. (1974) "Investigations with a Saprozoic Nematode, *Mesodiplogaster lheritieri,* on a Possible Respiratory Function of Air Swallowing," *Nematologica,* **20,** 52–60.

Stringfellow, F. (1974) "Hydroxyl Ion, an Attractant to the Male of *Pelodera strongyloides,*" *Proceedings of the Helminthological Society of Washington,* **41,** 4–10.

Ward, S. (1973) "Chemotaxis by the Nematode *Caenorhabditis elegans:* Identification of Attractants and Analysis of the Response by Use of Mutants," *Proceedings of the National Academy of Science, U.S.A.,* **70,** 817–821.

Chapter 5

Burrows, R. B. and Lillis, W. G. (1964) "The Whipworm as a Blood Sucker," *Journal of Parasitology,* **50,** 675–680.

Colam, B. J. (1971*a*) "Studies on Gut Ultrastructure and Digestive Physiology in *Rhabdias bufonis* and *R. sphaerocephala* (Nematoda: Rhabditida)," *Parasitology,* **62,** 247–258.

Colam, B. J. (1971*b*) "Studies on Gut Ultrastructure and Digestive Physiology in *Cosmocerca ornata* (Nematoda: Ascaridida)," *Parasitology,* **62,** 259–272.

Crofton, H. D. (1971) "Form Function and Behaviour," in *Plant Parasitic Nematodes,* Vol. I (ed. B. M. Zuckerman, B. M., Mai, W. F., and Rohde, R. A.) Academic Press, New York and London.

Croll, N. A. (1972) "Feeding and Lipid Synthesis in *Ancylostoma tubaeforme* Preinfective Larvae," *Parasitology,* **64,** 369–378.

Davey, K. G. (1964) "The Food of *Ascaris,*" *Canadian Journal of Zoology,* **42,** 1160–1162.

de Soyza, K. (1973) "Energetics of *Aphelenchus avenae* in Monoxenic Culture," *Proceedings of the Helminthological Society of Washington,* **40,** 1–10.

Doncaster, C. C. (1962) "Nematode Feeding Mechanisms, I. Observations on *Rhabditis* and *Pelodera,*" *Nematologica,* **8,** 313–320.

Doncaster, C. C. (1971) "Feeding in Plant Parasitic Nematodes: Mechanisms and Behaviour," in *Plant Parasitic Nematodes,* **II,** 137–157 (ed. Zuckerman, B. M., Mai, W. F., and Rohde, R. A.) Academic Press, New York and London.

Harpur, R. P. and Popkin, J. S. (1973) "Intestinal Fluid Transport: Studies with the Gut of *Ascaris lumbricoides,*" *Canadian Journal of Physiology and Pharmacology,* **51,** 79–90.

Jenkins, T. and Erasmus, D. A. (1969) "The Ultrastructure of the Intestinal Epithelium of *Metastrongylus* sp. (Nematoda: Strongyloidea)," *Parasitology,* **59,** 335–342.

Lee, D. L. (1968) "The Ultrastructure of the Alimentary Tract of the Skin-penetrating Larvae of *Nippostrongylus brasiliensis* (Nematoda)," *Journal of Zoology*, London, **154**, 9–18.

Lee, D. L. and Anya, A. O. (1968) "Studies on the Movement, the Cytology and the Associated Micro-organisms of the Intestine of *Aspiculuris tetraptera* (Nematoda)," *Journal of Zoology*, London, **156**, 9–14.

Read, C. P. (1966) "Nutrition of Intestinal Helminths," in *Biology of Parasites*, 101–126, (ed. Soulsby, E. J. L.) Academic Press, New York and London.

Read, C. P. (1968) "Some Aspects of Nutrition in Parasites," *American Zoologist*, **8**, 139–149.

Riding, I. L. (1970) "Microvilli on the Outside of a Nematode," *Nature*, London, **226**, 179–180.

Thorson, R. E. (1956) "Proteolytic Activity in Extracts of the Oesophagus of Adults of *Ancylostoma caninum* and the effect of Immune Serum on the Activity," *Journal of Parasitology*, **42**, 21–25.

Tietjen, J. H. and Lee, J. J. (1973) "Life History and Feeding Habits of the Marine Nematode *Chromadora macrolaimoides* Steiner," *Oecologia* (Berl.) **12**, 303–314.

Vanfleteren, J. R. (1974) "Nematode Growth Factor," *Nature*, London, **248**, 255–257.

Wells, H. S. (1931) "Observations on the Blood Sucking Activities of the Hookworm *Ancylostoma caninum*," *Journal of Parasitology*, **17**, 167–182.

Wright, K. A. (1974) "The Feeding Site and Probable Feeding Mechanism of the Parasitic Nematode *Capillaria hepatica* (Bancroft, 1893)," *Canadian Journal of Zoology*, **52**, 1215–1220.

Wyss, V. (1973) "Feeding of *Tylenchorhynchus dubius*," *Nematologica*, **19**, 125–136.

Zam, S. G. and Martin, W. E. (1969) "Binding of ^{60}Co-vitamin B_{12} by *Ascaris suum* Intestine," *Journal of Parasitology*, **55**, 480–485.

Chapter 6

Anderson, R. V. and Darling, H. H. (1964) "Embryology and Reproduction of *Ditylenchus destructor* Thorne, with emphasis on Gonad Development," *Proceedings of the Helminthological Society of Washington*, **31**, 240–256.

Anya, A. O. (1964) "Studies on the Structure of the Female Reproductive System and Eggshell Formation in *Aspiculuris tetraptera* Schulz (Nematoda Oxyuroidea)", *Parasitology*, **54**, 699–719.

Bird, A. F. (1968) "Changes Associated with Parasitism in Nematodes. III Ultrastructure of the Egg Shell, Larval Cuticle, and Contents of the Subventral Esophageal Glands of *Meloidogyne javanica*, with some observations on Hatching," *Journal of Parasitology*, **54**, 475–489.

Bird, A. F. (1971) *The Structure of Nematodes*, Academic Press, London and New York.

Bird, A. F. and Rogers, G. E. (1965) "Ultrastructure of the Cuticle and its Formation in *Meloidogyne javanica*", *Nematologica*, **11**, 224–230.

Bird, A. F. and Wallace, H. R. (1965) "The Influence of Temperature on *Meloidogyne hapla* and *H. javanica*," *Nematologica*, **11**, 581–589.

Boveri, T. (1899) "Die Entwicklung von *Ascaris megalocephala* mit besonder Rücksicht an die Kernverhältnisse," *Festschrift Kupfer*, pp. 283–340.

Bridge, J. (1974) "Hatching of *Tylenchorhynchus maximus* and *Merlinius icarus*", *Journal of Nematology*, **6**, 101–102.

Clarke, A. J. and Shepherd, A. H. (1966) "Picrolonic Acid as a Hatching Agent for the Potato Cyst Nematode *Heterodera rostochiensis* Woll," *Nature*, London, **211**, 546.

Clarke, A. J. and Shepherd, A. H. (1968) "Hatching Agents for the Potato Cyst Nematode *Heterodera rostochiensis* Woll," *Annals of Applied Biology*, **61**, 139–149.

Crofton, H. D. (1965) "Ecology and Biological Plasticity of Sheep Nematodes. I. The Effect of Temperature on the Hatching of Eggs of Some Nematode Parasites of Sheep," *Cornell Veterinarian*, **55**, 242–279.

Croll, N. A. (1974) "*Necator americanus*: Activity Patterns in the Egg and the Mechanism of Hatching," *Experimental Parasitology*, **35**, 80–85.

Davey, K. G. (1971) Moulting in a Parasitic Nematode, *Phocanema decipiens*. VI The Mode of Action of Insect Juvenile Hormone and Farnesyl Methyl Ether," *International Journal for Parasitology*, **1**, 61–66.

Davey, K. G. and Sommerville, R. I. (1974) "Moulting in a Parasitic Nematode, *Phocanema decipiens*. VII The Mode of Action of the Ecdysial Hormone," *International Journal for Parasitology*, **4**, 241–259.

Doncaster, C. C. and Shepherd, A. M. (1967) "The Behaviour of Second Stage *Heterodera rostochiensis* Larvae Leading to their Emergence from the Egg," *Nematologica*, **13**, 476–478.

Ellenby, C. (1968a) "Desiccation Survival of the Infective Larvae of *Haemonchus contortus*," *Journal of Experimental Biology*, **49**, 469–475.

Ellenby, C. (1968b) "On the Waterproofing Function of the Retained Second Stage Cuticle of the Third Stage Larva of *Haemonchus contortus*," *Parasitology*, **58**, 3.

Fairbairn, D. (1960) "Physiologic Aspects of Egg Hatching and Larval Exsheathment in Nematodes," in *Host Influence on Parasite Physiology*, (ed. L. A. Stauber), Rutgers University Press, New Brunswick.

Foor, W. E. and McMahon, J. T. (1973) "Role of the Glandular Vas Deferens in the Development of *Ascaris* Spermatozoa," *Journal of Parasitology*, **59**, 753–758.

Hertwig, O. (1890) "Vergleich der Ei und Samenbilslung bei Nematoden," *Archiv für Mikroskopische Antomie*, **36**, 1–38.

Lapage, G. (1935) "The Second Ecdysis of Infective Nematode Larvae," *Parasitology*, **27**, 186–206.

Lee, D. L. (1971) "The Structure and Development of the Spermatozoon of *Heterakis gallinarum* (Nematoda)," *Journal of the Zoological Society of London*, **164**, 181–187.

Ozerol, N. H. and Silverman, P. H. (1969) "Partial Characterization of *Haemonchus contortus* Exsheathing Fluid," *Journal of Parasitology*, **55**, 79–87.

Ozerol, N. H. and Silverman, P. H. (1970) "Further Characterization of Active Metabolites from Histotropic Larvae of *Haemonchus contortus* Cultured *in vitro*," *Journal of Parasitology*, **56**, 1119–1205.

Ozerol, N. H. and Silverman, P. H. (1972a) "Exsheathment Phenomenon in the Infective-stage Larvae of *Haemonchus contortus*," *Journal of Parasitology*, **58**, 34–44.

Ozerol, N. H. and Silverman, P. H. (1972b) "Enzymatic Studies on the Exsheathment of *Haemonchus contortus* Infective Larvae: the Role of Leucine Aminopeptidase," *Comparative Biochemistry and Physiology*, **42 B**, 109–121.

Pillai, J. K. and Taylor, D. P. (1968) "Biology of *Paroigolaimella bernesis* and *Fictor anchicoprophaga* (diplogasterinae) in Laboratory Culture," *Nematologica*, **14**, 159–170.

Rogers, W. P. (1958) "Physiology of the Hatching of Eggs of *Ascaris lumbricoides*," *Nature*, London, **181**, 1410–1411.

Rogers, W. P. (1960) "The Physiology of Infective Processes of Nematode Parasites; the Stimulus from the Animal Host," *Proceedings of the Royal Society* B, **152**, 367–386.

Rogers, W. P. (1962) *The Nature of Parasitism*, Academic Press, New York and London.

Rogers, W. P. (1965) "The Role of Leucine Aminopeptidase in the Moulting of Nematode Parasites," *Comparative Biochemistry and Physiology*, **14**, 311–321.

Rogers, W. P. (1966) "Exsheathing and Hatching Mechanisms in Helminths," in *Biology of Parasites* (ed. E. J. L. Soulsby)c, Academic Press, New York.

Rogers, W. P. (1970) "The Function of Leucine Aminopeptidase in Exsheathing Fluid," *Journal of Parasitology*, **56**, 138–143.

Rogers, W. P. and Sommerville, R. I. (1957) "Physiology of Exsheathment in Nematodes and its relation to Parasitism," *Nature*, London, **179**, 619–621.

Rogers, W. P. and Sommerville, R. I. (1960) "The Physiology of the Second Ecdysis of Parasitic Nematodes," *Parasitology*, **50**, 329–348.

Shepherd, A. H. (1962) *The Emergence of Larvae from Cysts in the Genus Heterodera*. Commonwealth Agricultural Bureau, Farnham Royal, England.

Sommerville, R. I. (1954) "The Second Ecdysis of Infective Nematode Larvae," *Nature, London*, **174**, 751–752.

Von Beneden, E. (1883) "Recherches sur la fécondation et la maturation," *Archives de Biologie*, **4**, 265–641.

Wallace, H. R. (1966) "The Influence of Moisture Stress on the Development, Hatch and Survival of Eggs of *Meloidogyne javanica*," *Nematologica*, **12**, 57–69.

Whitlock, J H. (1971) "Ecdysis of *Haemonchus* and Hypotheses," *Cornell Veterinarian*, **61**, 349–361.

Wilson, P. A. G. (1958) "The Effect of Weak Electrolyte Solutions on Hatching Rate of *Trichostrongylus retortaeformis* (Zeder) and its Interpretation in Terms of a Proposed Hatching Mechanism of Strongylid Eggs," *Journal of Experimental Biology*, **35**, 584–601.

Zur Strassen, O. (1896) "Embryonalentwicklung der *Ascaris megalocephala*," *Archiv Entwickburgsmechanik*, **3**, 27–105; 132–190.

Chapter 7

Anderson, R. C. (1957) "The Life Cycles of Dipetalonematid Nematodes (Filarioidea, Dipetalonematidae): the Problem of their Evolution," *Journal of Helminthology*, **31**, 204–224.

Goodey, T. (1930) "On a Remarkable New Nematode, *Tylenchinema oscinellae* Gen. et sp.n. Parasite in the Fruit Fly *Oscinella frit* L., Attacking Oats," *Philosophical Transactions of the Royal Society of London B*, **218**, 315–343.

Heyneman, D. (1966) "The Life Cycles of Nematodes Parasitic in Man: an Evolutionary Sequence," *Medical Journal of Malaya*, **20**, 249–263.

Jenkins, T. (1970) "A Morphological and Histochemical Study of *Trichuris suis* (Schrank, 1788) with special reference to the Host-parasite Relationship," *Parasitology*, **61**, 357–374.

Klingler, J. (1970) "The Reaction of *Aphelenchoides fragariae* to Slit-like Micro-openings and to Stomatal Diffusion Gases," *Nematologica*, **16**, 417–422.

Sprent, J. F. A. (1962) "The Evolution of the Ascaridoidea," *Journal of Parasitology*, **48**, 818–824.

Chapter 8

Ayala, A. and Ramirez, C. T. (1964) "Host Range, Distribution and Bibliography of the Reniform Nematode *Rotylenchulus reniformis*, with special reference to Puerto Rica," *Journal of Agriculture of the University of Puerto Rica*, **48**, 140–161.

Beaver, P. C. (1945) "Immunity to *Necator americanus* Infection," *Journal of Parasitology*, **31**, Suppl. 18.

Hawking, F. (1967) "The 24-hour Periodicity of Microfilariae: Biological Mechanisms Responsible for its Production and Control," *Proceedings of the Royal Society B*, **169**, 59–76.

Jarrett, W. F. H. (1966) "Pathogenic and Expulsive Mechanisms in Gastro-intestinal Nematodes," in *Pathology of Parasitic Diseases* (ed. A. E. R. Taylor), Fourth Symposium of the British Society for Parasitology, Blackwell Scientific Publications, Oxford.

Jenkins, D. C. & Phillipson, R. F. (1972) "Evidence that the Nematode *Nippostrongylus brasiliensis* can Adapt and Overcome the Effects of Host Immunity," *International Journal for Parasitology*, **2**, 353–359.

Jones, M. G. K. & Northcote, D. H. (1972) "Nematode-induced Syncytium—a Multinucleate Transfer Cell," *Journal of Cell Science*, **10**, 789–809.

Love, R. J. (1975) "*Nippostrongylus brasiliensis* Infection in Rats," *International Archives of Allergy and Applied Immunology*, **48**, 211–219.

Maplestone, P. A. (1933) "Creeping Eruption produced by Hookworm Larvae," *Indian Medical Gazette*, **68**, 251–257.

Matthews, B. E. & Croll, N. A. (1974) "Comparative Stereoscan Electron Micrographs of Nematode Heads," *Journal of Nematology*, **6**, 131–134.

Ogilvie, B. M. (1974) "Antigenic Variation in the Nematode *Nippostrongylus brasiliensis*," in *Parasites in the Immunized Host: Mechanisms of Survival*, Ciba Foundation Symposium 25 (new series).

Phillipson, R. F. (1974) "Intermittent Egg Release by *Aspiculuris tetraptera* in Mice," *Parasitology*, **69**, 207–213.

Stoll, N. R. (1929*a*) "The Occurrence of Self-cure and Protection in Typical Nematode Parasitism," *Journal of Parasitology*, **15**, 147–148.

Stoll, N. R. (1929*b*) "Studies with the Stongylid Nematode, *Haemonchus contortus*. I. Acquired Resistance of Hosts under Natural Reinfection Conditions out of Doors," *American Journal of Hygiene*, **10**, 384–418.

Trudgill, D. L. (1967) "The Effect of Environment on Sex Determination in *Heterodera rostochiensis*," *Nematologica*, **13**, 263–272.

Chapter 9

Armour, J. and Bruce, R. G. (1974) "Inhibited Development in *Ostertagia ostertagi* Infections—a Diapause Phenomenon in a Nematode," *Parasitology*, **69**, 161–174.

Barrett, J. (1969) "The Effect of Ageing on the Metabolism of the Infective Larvae of *Strongyloides ratti*, Sandground, 1925," *Parasitology*, **59**, 3–17.

Bhatt, B. D. and Rohde, R. A. (1970) "The Influence of Environmental Factors on the Respiration of Plant-parasitic Nematodes," *Journal of Nematology*, **2**, 277–285.

Blitz, N. H. & Gibbs, H. C. (1972*a*) "Studies on the Arrested Development of *Haemonchus contortus* in sheep. I. The Induction of Arrested Development," *International Journal for Parasitology*, **2**, 5–12.

Blitz, N. H. & Gibbs, H. C. (1972*b*) "Studies on the Arrested Development of *Haemonchus contortus* in sheep. II Termination of Arrested Development and the Spring Rise Phenomenon," *International Journal for Parasitology*, **2**, 13–22.

Cooper, A. F. & Van Gundy, S. D. (1971) "Senescence, Quiescence and Cryptobiosis," in *Plant Parasitic Nematodes*, Vol. II, (ed. Zuckerman, B.M., Mai, W. F. and Rohde, R. A.), Academic Press, London and New York.

Costello, L. C. and Grollman, S. (1958) "Oxygen Requirement of *Strongyloides papillosus* Infective Larvae," *Experimental Parasitology*, **7**, 319–327.

Croll, N. A. & Matthews, B. E. (1973) "Activity, Ageing and Penetration of Hookworm Larvae," *Parasitology*, **66**, 279–289.

Ellenby, C. (1969) "Dormancy and Survival in Nematodes," *Symposia of the Society of Experimental Biology*, **23**, 83–97.

Elliott, A. (1954) "Relationship of Ageing, Food Reserves and Infectivity of Larvae of *Ascaridia galli*," *Experimental Parasitology*, **3**, 307–320.

Feder, W. A. and Feldmesser, J. (1955) "Further Studies on Plant Parasitic Nematodes Maintained in Altered Oxygen Tensions, " *Journal of Parasitology*, **41**, Suppl. 47.

Fernando, M. A., Stockdal, P. H. G. and Ashton, G. C. (1971) "Factors Contributing to the Retardation of Development of *Obeliscoides cuniculi* in Rabbits," *Parasitology*, **63**, 21–29.

Gershon, H. and Gershon, D. (1970) "Detection of Inactive Enzyme Molecules in Ageing Organisms," *Nature*, London, **227**, 1214–1217.

Matthews, B. E. (1972) "Invasion of Skin by Larvae of the Cat Hookworm, *Ancylostoma tubaeforme*," *Parasitology*, **65**, 457–467.

Matthews, B. E. (1975) "Mechanism of Skin Penetration by *Ancylostoma tubaeforme* Larvae," *Parasitology*, **70**, 25–38.

Michel, J. F. (1974) "Arrested Development of Nematodes and Some Related Phenomena," *Advances in Parasitology*. **12**, 279–366. (ed. B. Dawes), Academic Press, London and New York.

Norris, D. E. (1971) "The Migratory Behaviour of the Infective Stage Larvae of *Ancylostoma braziliense* and *Ancylostoma tubaeforme* in Rodent Paratenic Hosts." *Journal of Parasitology*, **57**, 998–1009.

Olsen, W. O. and Lyons, E. T. (1965) "Life Cycle of *Uncinaria lucasi* Stiles, 1901 (Nematoda: Ancylostomatidae) of Fur Seals, *Callorhinus ursinus* Linn., on the Pribilof Islands, Alaska," *Journal of Parasitology*, **51**, 689–700.

Rogers, W. P. (1939) "The Physiological Ageing of Ancylostome Larvae," *Journal of Helminthology*, **17**, 195–202.

Rogers, W. P. (1940) "The Physiological Ageing of the Infective Larvae of *Haemonchus contortus*," *Journal of Helminthology*, **18**, 183–192.

Rose, J. H. (1963) "Observations on the Free-living Stages of the Stomach Worm *Haemonchus contortus*," *Parasitology*, **53**, 469–481.

Schad, G. A., Chowdbury, A. B., Dean, C. G., Kochar, V. K., Nawalinski, T. A., Thomas, J. and Tonascia, J. A. (1973) "Arrested Development in Human Hookworm Infections: an Adaptation to a Seasonally Unfavourable External Environment," *Science*, **180**, 502–504.

Van Gundy, S. D., Bird, A. F. and Wallace, H. R. (1967) "Ageing and Starvation of Larvae in *Meloidogyne javanica* and *Tylenchus semipenetrans*," *Phytopathology*, **57**, 559–571.

Chapter 10

Broome, A. W. J., (1961) "Studies on the Mode of Action of Methyridine," *British Journal of Pharmacology*, **17**, 327–335.

Canning, E. U., (1973) "Protozoal Parasites as Agents for Biological Control of Plant-parasitic Nematodes," *Nematologica*, **19**, 342–348.

del Castillo, J. and Morales, T. (1969) "Electrophysiological Experiments in *Ascaris lumbricoides*," in *Experiments in Physiology and Biochemistry*, II, 209–273. (Ed. Kerkut, G. A.) Academic Press, New York and London.

Desowitz, R. S., (1971) "Antiparasite Chemotherapy," *Annual Review of Pharmacology*, **11**, 351–367.

Duddington, C. L. (1956) "The Predacious Fungi: Zoopagales and Moniliales," *Biological Reviews*, **31**, 152–193.

Evans, A. A. F. (1973) "Mode of Action of Nematicides," *Annals of Applied Biology*, **75**, 439–479.

Kates, K. C., Colglazier, M. L. and Enzie, F. D. (1973) "Experimental Development of a Cambendazole Resistant Strain of *Haemonchus contortus* in Sheep," *Journal of Parasitology*, **59**, 169–174.

Lee, C. C. and Miller, J. H. (1969) "Fine Structure of the Intestinal Epithelium of *Dirofilaria immitis* and Changes Occurring after Vermicidal Treatment with Caparsolate Sodium," *Journal of Parasitology*, **55**, 1035–1045.

Nelmes, A. J. (1972) "Behavioural Responses of *Heterodera rostochiensis* Larvae to Aldicarb and its Sulphoxide and Sulfone," *Journal of Nematology*, **2**, 223–227.

Nelson, G. S. (1972) "Human Behaviour in the Transmission of Parasitic Disease," in *Behavioural Aspects of Parasite Transmission*, (ed. Canning, E. U., and Wright, C. A.) 109–122, Academic Press, London and New York.

Pitts, N. E. and Migliardi, J. R. (1974) "Antiminth *(Pyrantel pamoate),*" *Clinical Pediatrics*, **13**, 87–94.

Poynter, D. (1963) "Parasitic Bronchitis," *Advances in Parasitology 1*, 179–212 (ed. Ben Dawes), Academic Press, London and New York.

Prichard, (1970) "Mode of Action of the Anthelminthic Thiabendazole in *Haemonchus contortus*," *Nature*, **228**, 684.

Sayre, R. M. and Keeley, L. S. (1969) "Factors Influencing *Catenaria anguillulae* Infections of a Free-living and Plant-parasitic Nematode," *Nematologica*, **15**, 492–502.

Saz. H. J. and Bueding, E. (1966) "Relationships between Anthelminthic Effects and Biochemical and Physiological Mechanisms," *Pharmacological Reviews*, **18**, 871–894.

Standen, O. D. (1963) "Chemotherapy of Helminthic Infections," in *Experimental Chemotherapy*, Vol. 1, 701–892, (ed. Schnitzer, R. J. and Hawking, F.), Academic Press, New York and London.

Viglierchio, D. R. and Croll, N. A. (1969) "The Comparative Effect of Chloramines on a Range of Nematodes," *Journal of Nematology*, **1**, 35–39.

Yeates, G. W. (1969) "Predation by *Mononchoides potohikus* (Nematoda: Diplogasteridae) in Laboratory Culture," *Nematologica*, **15**, 1–9.

Index

abomasum 118
absorption 84
acanthocephalans 4, 38
acclimatization 75, 76
acetylcholine 20, 40, 170
acorn worms 3
Acrobeles 108, 177
action potential 37
activity 54, 155, 172
Acuaria 9
Aedes togoi 30
aerobiosis 24, 164–165
air breathing 55, 64
Al Hadithi, I. 55
aldicarb 172
alkyl halides 176
amino acids 84, 88
AMP 69
amphids 43, 53
anaerobiosis 24, 164–165
Ancylostoma 8, 94, 124
A. brasiliensis 146, 156
A. caninum 26, 86, 89, 90, 102, 108, 144, 168
A. ceylanicum 144
A. duodenale 26, 79, 89, 146, 157, 175
A. tubaeforme 26, 57, 58, 73, 74, 75, 96, 97, 101, 144, 145, 155, 156
Anderson, R.C. 128
Anderson, R.V. 100
Angiostrongylus cantonensis 164
Anguina agrostis 104
A. tritici 161, 162
anions 69
Anisakis 8, 145
Annelida 4
anthelminthics 20, 40, 168–177
antienzymes 139
anus 65, 79
Anya, A.O. 41, 88, 103, 186

Aphelenchoides blastophthorus 10, 27, 179
A. composticola 27, 30
A. fragariae 69, 70
A. obsistus 165
A. ritzemabosi 139, 161
Aphelenchus avenae 10, 18, 22-27, 61, 95–97, 164, 165, 176
Aphodius 121, 122
appendix 143
Aristotle 3
Armigeres subalbatus 30
Arthrobotrys oligospora 181
Arthropods 4
Ascaridia galli 9, 86, 90 113, 152–154
Ascaris lumbricoides 7, 8, 20, 22, 25, 26, 31, 32, 37, 56, 65, 66, 70, 75, 79, 82, 84–86, 98–104, 108, 113, 125, 126, 142–145, 153, 170, 174, 175
A. megalocephala 105
A. suum 20, 86
Aschelminthes 6
Aspiculuris tetraptera 9, 26, 28, 41, 82, 88, 99, 136
Aspidodera 9
attractants 69
auxin 93
axenic culture 12
Ayala, A. 148

bacterial feeding 94
Barrett, J. 154
Bdellonyssus bacoti 28
Beaver, P. 146
Beguet, B. 13
behaviour 12, 54–78
behavioural mutants 12
Belonolaimus gracilis 165
bephenium 169, 171
bilateralism 6

biological control   179-183
Bird, A.F.   2, 33, 111, 112, 116
black head disease   139
Blair, A.   58
Blitz, N.H.   158
blood feeding   89-91
Bodo caudatus   179, 180
Botrytis cinerea   96
Boveri, T.   106
Bradynema   86-87
Breinlia booliati   9, 26-28, 30
B. sergenti   30
Brenner, S.   12
Bridge, J.   112
Bruce, R.G.   158
Brugia malayi   9, 26, 28, 168, 173
B. pahangi   26, 28
Brun, J.L.   13, 16
Bunostomum   8, 141
Burrows, R.B.   91
bursa   101

Caenorhabditis   8, 24, 80, 164, 165
C. briggsae   17, 20, 21, 88
C. elegans   13, 14, 16, 17, 20, 44, 57, 65, 66, 68, 69, 78, 82, 104, 135
Callorhinus ursinus   157
calories   95-96
Camallanus   9, 74
cambendazole   171
Canning, E.U.   179-181
caparsolate sodium   174
Capillaria hepatica   10, 50, 52, 91, 142
carbohydrate metabolism   22
carbon dioxide   65, 69-71, 120
Catenaria anguillulae   182, 183
cations   69
caudal papillae   43, 44
Celsus   3
cephalic setae   43, 44
Cephalobus   64
cercariae   125
Chabertia   141
Chaetagnatha   4
chemosensitivity   68-73
chemotherapy   167-177
Chitwood, B.G.   2
Chitwood, M.B.   2
chloramine   171, 176-177
Chlorella   3
chordata   4
Chromadora   99
C. macrolaimoides   94

Chromadorina bioculata   10, 44, 45, 47, 49, 74
chromatrope   48, 74
Chrysops   137
cilia   6
cinematography   54
Cobb, N.A.   2, 183
Coelenterata   4
Colam, B.J.   90
Cooper, A.F. Jr.   165
Cooperia   141
copulation   51, 53, 55, 100
coracidium   125
Corynebacterium fascians   139
Cosmocera ornata   82
Costello, L.C.   155
creeping eruption   146
Crick, F.   12
Criconemoides   177
Crofton, H.D.   2, 31, 32, 62, 87-88
Croll, N.A.   41, 48, 49, 54, 55, 63, 68, 72, 74-77, 96, 97, 111, 156, 177
cryptobiosis   24
Ctenophora   4
Culex fatigans   137
culture of nematodes   22-26, 88
cutaneous larval migrans   146
cuticle   33, 36, 86-87, 116, 117
cuticular projections   47

Dactylaria candida   181
Darling, H.H.   100
dauer larvae   122
Davey, K.G.   116
D-D   176
defaecation   55, 65
del Castillo, J.C.   40
Deontostoma timmerchioi   51, 53
dermal light sense   49, 72
Desmodera   10
Desmoscoleicida   47
Desowitz, R.S.   173
de Soyza, K.   95-96
determinant cleavage   106-108
development of nematodes   98-119
developmental biology   12
developmental inhibitors   19
Dictyocaulus filaria   8, 168
D. viviparus   142, 148, 168, 178
diethylcarbamazine   169, 173
Dioctophyme renale   7, 10, 142
Dipetalonema witeae   26, 28, 44
Diphyllobothrium latum   86

Dirofilaria 128, 130
D. immitis 174
disease 138
dithiazanine 169, 171, 173
Ditylenchus destructor 10, 27, 76, 92, 100
D. dipsaci 27, 69, 70, 73, 75, 76, 161, 162, 164, 178, 179, 180
D. myceliophagus 27, 92
D. triformis 161
Dolichodorus heterocephalus 165
Dolichosaccus heterocephalus 165
Doncaster, C.C. 54, 59, 61, 81, 92, 112
Dorylaimus 177
Dracunculus medinensis 1, 7, 9, 104, 139, 166–167
Drosophila melanogaster 12, 16
drugs, 40, 167
Duboscquia penetrans 180–181
Duddington, C.L. 181

Echinodermata 4
EDB 176
eelworms 3
eggs 15, 58, 59, 60, 102–103, 111, 125
electron microscope 2
Ellenby, C. 117, 161–162
Elliot, A. 152
embryology 104–108
Enoplus communis 10, 45, 46, 135
Enterobius vermicularis 9, 20, 26, 82, 126, 127, 143–145, 170, 174, 175
Entoprocta 4
enzyme secretion 89, 93, 94, 111
epinephrine 41
Escherichia coli 3
Eustrongylides ignotus 10, 156
eutely 17
Evans, A.A.F. 169
everted sacs 84
exsheathment 55, 115–119

faecal production 136
Fairbairn, D. 114–115
fecundity 104
feeding 55, 61, 62, 63, 79–97
feeding stimulus 90
Fictor anchicoprophaga 109
fiery serpent 1
Filaria 9
filariasis 28
flagellates 179
flow diagrams 63
flukes 3

Foor, W.E. 100
frit fly 134

galvanotaxis 77
gamma-amino butyric acid 41
ganglia 37–41
Gastrotricha 4, 38
genetics 13
geosensitivity 77
gerontology 17–19
Gershon, D. 17–19, 156
giant cell 93, 140
glycogen 22, 24, 165
Gnathostoma 9
Gongylonema 9
Gordian worms 5
Green, C.D. 71, 72
Greet, D.N. 70
growth curves 96
gubernaculum 100
Guinea worm 1, 166–167

Haemonchus contortus 8, 26, 41, 69, 90, 118–119, 141, 147–9, 153, 154, 156–159, 162, 176
haloxon 173
Hammerschmidtiella 9
Harpur, R.P. 85
Harris, J.E. 31, 32
hatching 55, 58, 59, 108, 109–115
hatching fluid 113
Hawking, F. 137
Hayalella 129
Heligmosomum 8
Helix aspersa 41
helminthology 2, 3
Hemicriconemoides chitwoodi 108
hemizonid 43, 52
hepatocytes 91
Hertwig, O. 105
Heterakis gallinarum 9, 51, 98, 139
Heterodera 7, 10, 27, 51, 72, 92, 99, 104, 113, 115, 139, 161
H. avenae 115
H. rostochiensis 20, 59, 61, 71, 104, 109, 112, 115, 161, 165, 172
H. schachtii 77, 115
Heyneman, D. 124
histamine 149
Histomonas meleagridis 139
hookworms 3, 11, 64, 66, 89, 144–145, 157
Hope, W.D. 51, 53

INDEX

husk 178
Hydromermis 140
hydrostatic pressure 6, 31, 32, 65
Hyman, L.H. 6
hypodermis 36

immunity 27, 149–150
immunosupressants 25
individuality 58
indole acetic acid 93
infective larvae 96–97, 122–124, 138, 154–156, 153–165
infectivity 152
ingestion rates 95
integrated control 184
intestine 36, 62, 65, 79, 85, 90
Iotonchus 80, 183

Jarrett, W.F.H. 147
Jenkins, D.C. 150
Jenkins, T. 126
jird 28
Jones, M.G.K. 140
juvenile 6

Kinorhyncha 4
Kisiel, M. 19
Klingler, J. 65, 69, 120
klinokinesis 72
Kunz, P. 65

Lapage, G. 117
larval stages 6
leaping behaviour 55, 66
Lee, D.L. 2, 40, 88
Legerella helminthosporum 179–180
Leidynema 9
leucine aminopeptidase 119
life cycles 1, 120–134
Lillis, W.G. 91
Linnaeus, C.V. 3
lipid metabolism 22, 24, 85, 96, 155
Litomosoides carinii 9, 26–28, 29, 77, 168, 173
Loa loa 26, 137
longevity 154, 156
Longidorus 10, 108, 139
Love, R.J. 151
lung 142

mamelons 43
Manson, Sir Patrick 1
Maplestone, P.A. 146

Matthews, B.E. 144, 154, 155
McLaren, D.J. 40
mechanosensitivity 75
Megaselia halterata 86
melanin 47
Meloidogyne 10, 27, 92, 93, 102, 104, 139, 165, 179
M. hapla 79, 110
M. icognita 93
M. javanica 104, 110, 112, 156, 164, 180
M. nassi 108
meniscus 66, 67
Meriones libycus 28
Mermis nigrescens 10, 48, 74, 79, 102, 104
Mesodiplogaster lheritieri 64
metacorpus 62
metamerism 6
Metastrongylus 8, 148
methyridine 174
Metoncholaimus 64
Michel, J.F. 160
microbrowsers 139
microfilaria 130–132, 137
microvilli 87
migration 55
miracidia 125
model nematodes 3, 12, 18–30
molluscs 4
Monohystera 10
Mononchus 10, 80, 183
M. composticola 179, 180
moulting 55, 96, 115–119
movement 19, 34, 40, 41, 52, 54
muscles 6, 34, 36
myofilaments 35

Necator americanus 8, 26, 59, 60, 89, 104, 108, 111, 112, 144, 146, 175
Nelmes, A.J. 172
Nelson, G.S. 168
nematicides 168
Nematodirus battus 8, 111, 118, 141, 159, 160
N. fillicollis 164
nematology 2
Nematomorpha 4
Nematospiroides dubius 28, 70, 101, 168
Nemertina 4
nerve ring 35–41
nerves 35–41
neuroanatomy 37–41, 78
neuromuscular physiology 2, 6, 35, 40, 56
nictating 55, 66

Nippostrongylus brasiliensis 8, 27, 51, 74, 79, 83, 86, 90, 94, 99, 149–151, 156, 168
nodule 132
norepinephrine 41
Norris, D.E. 156
Northcote, D.H. 140
Nosema eurytremae 180

Obeliscoides cuniculi 158
ocellus 43, 45–49, 53
oesophageal pumping 59, 162
Oesophagostomum colubrianum 118, 141
oesophagus, see pharynx
Oil Red O 97
Olsen, W.O. 157
Onchocerca volvulus 9, 26, 28, 128, 130–132, 137, 152, 166
onchomiracidium 125
Onychophora 4
opisthodelphic 102
orientation responses 68
Ornithodirus tartakovskyi 28
Oscinella frit 134
osmotic stress 111
Ostertagia circumcincta 118, 119, 141, 147, 176
O. ostertagi 147, 158
ovary 102
oviposition 55
oxygen relations 22, 24, 25, 165
oxyhaemoglobin 48
Oxyspirura 129, 130
Ozerol, N.H. 119

Panagrellus 8, 64
P. redivivus 13, 17–20, 67, 135, 182
P. silusiae 13, 17–20
Panagrolaimus rigidus 70, 71
Parafilaria multipapillosa 129, 131
Parascaris equorum 8, 98, 103, 104, 105, 108
Paroigalaimella bernensis 109
pathology 138
patterned behaviour 63
Pelodera 64, 80
P. coarctata 121, 122
P. strongyloides 72, 77, 122–123
P. teres 104
penetration 55, 64, 123, 124, 145, 155
pepsin 147
periodicity 136, 157

pharynx 61–63, 79, 81
phasmids 43, 44, 53
phenothiazine 171
Phillipson, R.F. 136
Philodina citrina 16
Phocanema decipiens 41, 116
photoklinokinesis 72–73
photoorthokinesis 73
photosensitivity 73–74
Physaloptera 9
pigment 49, 90
pinworms 3, 126, 127
piperazine 20, 40, 168, 171
Placentonema gigantissima 7
Platyhelminthes 4
Plectus 10, 164
Poiseuilles' Law 81
Porifera 4
Porrocaecum 8
Pratylenchus penetrans 108, 140, 161, 164, 179
predatory nematodes 80
Priapulida 4
Prionchulus 80, 104, 183
procorpus 62
proctodaeum 6
prodelphic 102
Proleptus 9
proprioception 59
Protostrongylus 8
Protozoa 179–180
Pseudalius 8
pseudocoelom 6, 32, 35, 50, 84
pseudocoelomic fluid 32, 84
pyrvinium 169–171

Radapholus similis 165
Rana temporaria 82
Read, C.P. 84, 91
rectum 65, 79
Reesimermis 10
Remane, A. 6
respiratory quotient 155
Rhabdias bufonis 8, 74, 90, 94, 122
R. nigrovenosa 90
R. sphaerocephala 90
rhabditiform 125
Rhabditis 8, 64, 77, 80, 99, 102, 165
R. oxycerca 71
R. teres 108
Rhabdochona ovifilamenta 129
rhabdomere 47
Rogers, W.P. 111, 113, 114, 119, 154

root diffusates 115
Rotifera 4
Rotylenchulus reniformis 148
roundworms 3
r-strategies 135

Samoiloff, M.R. 13
Sayre, R.M. 182–183
Schaad, G.A. 157
seals 157
segmentation 6, 80
Seinura 183
self cure 148, 159
seminal vesicle 99
senescence 16
sense organs 42–51
serotonin 41
sex attraction 70–72
Seymour, M.K. 59, 61
sheath 182
shock reactions 55
Sigmodon hispidus 28, 29
Silverman, P.H. 119
Simulium 28, 131, 140, 141, 166
Sipunculida 4
site specificity 141
size range 7
Smith, J.M. 48, 49, 75
somatic setae 43, 44, 48
Sommerville, I.R. 116, 118, 119
spermatogonia 100
spermatozoa 99–100
spicule 43, 51, 100
Spirura 9
Sprent, J.F.A. 126
spring rise 158
Stephanofilaria 130–131
stimulus response 55
Stoll, N.R. 148
stomodaeum 6, 62, 144
Stringfellow, F. 72
Strongyloides 8, 123, 141
S. papillosus 156, 168
S. ratti 26, 28, 73, 74, 155, 168
S. stercoralis 26, 28, 104, 123, 174
Strongylus edentatus 8, 164
S. vulgaris 86, 164
stylet 59, 61, 63, 112, 139, 172
Subulura 9
swarming 55, 66
symmetry 53
sympathetic nervous system 28, 92
synchronized cultures 18

syncytia 93, 140
Syngamus trachea 8, 100
Syphacea 9, 26
S. obvelata 28

tapeworms 3
Tardigrada 4
tartar emetic 171
telogonic 98
testis 98–99
tetrachlorethylene 171
Tetradonema 10
Tetrameres 7, 9
tetramisole 175
Thelazia 9, 129, 130
thermotaxis 64, 74, 75
thiabendazole 20, 21, 171, 175
thiamine 41
Thoracostoma californicum 44, 45
Thorson, R.E. 89
Tietjen, J.H. 94
Toxascaris cati 8, 159
toxicity 20
Toxocara canis 8, 86, 113, 136, 145, 146, 159, 160
trachea 124
transmission 1
transport, amino acids 84
  glucose 84–88
  phosphate 85–88
  vitamins 86
  water 85–88
trichina worm 3
Trichinella spiralis 1, 10, 25–28, 126–128, 138–9, 145, 152, 156, 166–167
Trichodorus christei 164
Trichonema 8, 72
Trichostrongylus 8, 26, 108, 141, 149
T. axei 118, 156
T. colubriformis 73, 118, 146, 176
T. retortaeformis 26, 77, 111
Trichuris bovis 10, 26
T. muris 26, 28, 168
T. ovis 26, 141
T. suis 91
T. trichiura 28, 91, 108, 126, 127, 174
T. vulpis 86, 91, 168, 174
Tripius 10
Tripyla 177
Trudgill, D.L. 140
Turbatrix aceti 8, 17, 18, 20, 21, 78, 156
Turbellaria 4
turgor, see hydrostatic pressure

Tylenchinema oscinellae 134
Tylenchorhynchus claytoni 161
T. dubius 92
T. icarus 73
T. maximus 112
Tylenchulus semipenetrans 165

Uncinaria lucasi 157
U. stenocephala 168
uterus 102

vaccination 178
Vanfleteren, J.R. 88
Van Gundy, S.D. 165
vas deferens 99
ventro-lateral supplement 51, 53
Viglierchio, D.R. 177
vinegar eelworm, see *Turbatrix aceti*
visceral larval migrans 146, 160
von Beneden, E. 105
vulva 14, 23, 25, 102

Wallace, H.D. 2

Ward, S. 13, 68
Watson, J.M. 12
wave movements 56
Webster, J.M. 93
Wells, H.S. 89
whipworms 91
Whitlock, J.H. 119
Wilson, P.A.G. 111
worms 3
Wright, K.A. 50, 52
Wuchereria bancrofti 9, 26, 28, 137, 152, 173
Wyss, U. 92

Xiphinema 10, 108, 135, 139
x-irradiated larvae 178

Yeates, G.W. 183

zoology of nematodes 4–7
Zuckerman, B.M. 19